Columbia University

Contributions to Education

Teachers College Series

No. 340

AMS PRESS
NEW YORK

# CONSUMERS USES OF ARITHMETIC

AN INVESTIGATION TO DETERMINE THE ACTUAL USES MADE OF ARITHMETIC IN ADULT SOCIAL LIFE, EXCLUSIVE OF VOCATIONAL USES

By

A. O. BOWDEN, Ph.D.

Teachers College, Columbia University
Contributions to Education, No. 340

Bureau of Publications
Teachers College, Columbia University
NEW YORK CITY
1929

**Library of Congress Cataloging in Publication Data**

Bowden, Aberdeen Orlando, 1881-1946.
Consumers uses of arithmetic.

Reprint of the 1929 ed., issued in series: Teachers College, Columbia University. Contributions to education, no. 340.
Originally presented as the author's thesis, Columbia.
Bibliography: p.
1. Arithmetic. 2. Arithmetic--Study and teaching. I. Title. II. Series: Columbia University. Teachers College. Contributions to education, no. 340.
QA142.B6 1972 375'.513 71-176583
ISBN 0-404-55340-0

Reprinted by Special Arrangement with Teachers College Press, New York, New York

From the edition of 1929, New York
First AMS edition published in 1972
Manufactured in the United States

AMS PRESS, INC.
NEW YORK, N. Y. 10003

## ACKNOWLEDGMENTS

More than thirteen hundred and sixty persons have coöperated in the present study. It is obviously impossible in this short note to render to each one individually my thanks and appreciation. To all who have in any way given their time and attention to the collection and tabulation of the data which are contained in the following pages I wish to express my sincere thanks.

First of all I wish to acknowledge my indebtedness to Superintendents Herbert H. Howes, Norwood, Mass.; C. F. Capps, Tupelo, Miss.; A. H. Hughey, El Paso, Texas; George Morriss, Bloomfield, N. J.; Frank Appell, Portsmouth, Ohio; R. W. Kirk, Oregon City, Ore.; and to their Sixth Grade teachers for their unselfish assistance in distributing and collecting the blanks represented by Appendix B. Also my thanks are due to Superintendent R. D. Bowden, Glenn Ellyn, Ill., and to Professor R. A. Edwards, Principal of the Training School of Eastern State Teachers College, Richmond, Ky., for data from Appendix A.

To Professor D. H. Kulp II of Teachers College, Columbia University, who has given me much encouragement in his sympathetic and valuable suggestions, I am deeply indebted. To Professors William A. McCall, Milo B. Hillegas and J. Ralph McGaughy of Teachers College much appreciation is hereby expressed for their generous assistance in the initial stages of this study and for their continued valuable guidance throughout the investigation. Professor McGaughy has been especially helpful in his suggestions in the statistical part of the text.

My greatest debt is due to Professor David Snedden, who by his keen insight into the problem, his encouragement and patience in correcting mistakes in methods of procedure is largely responsible for the success of this undertaking. Under his guidance the whole task has been an inspiration and a joy. Those who are familiar with his writing and class lectures will be able to recognize many of his ideas throughout these pages. A deep sense of obligation is here expressed to Roy S. West for mimeographing and other types of clerical assistance.

To Katharine M. Bowden an unredeemable debt is acknowledged for tabulations of certain parts of the data and also for sacrifices in other ways, too numerous to mention here, which have made the study possible.

A. O. Bowden

# CONTENTS


CHAPTER PAGE

I. INTRODUCTION ..... 1

i. A Statement of the Problem ..... 1
ii. Curriculum Making—a Continuous Process ..... 2
iii. Two Views in Curriculum Building ..... 4
iv. The Child Develops in Adjusting to Adult Environment.. 4
v. Culture—a Process of Growth and Discard ..... 5
vi. Previous Methods of Curriculum Revision ..... 6
vii. What is Valuable? ..... 8

II. HISTORICAL SKETCH OF INVESTIGATIONS TOUCHING ON THE PRESENT INVESTIGATIONS ..... 9

STAGES OF DEVELOPMENT:
i. Stage of Accretion ..... 10
ii. Pedagogical Period ..... 11
iii. Administrator's Opinion Period ..... 11
iv. Comparative Period ..... 13
v. Intelligence Test Period ..... 14
vi. Business Man's Opinion Period ..... 15
vii. Sociological Period ..... 15
viii. Brief Summary and Criticism ..... 16
ix. Bibliography ..... 18

III. METHOD USED IN PRESENT STUDY ..... 21

i. Method in General ..... 21
ii. Studies Based Upon Opinion and Estimate ..... 22
iii. Remembered Use—the Basis of this Study ..... 23
iv. Method Used in Solution of this Problem ..... 25
v. Analysis and Description of Method Followed ..... 26
vi. Method Used in This Investigation compared with Methods Used in Previous Investigations ..... 30

IV. PRESENTATION OF THE DATA ..... 34

V. CENTRAL TENDENCIES IN THE DATA ..... 36

i. Comments on Central Tendencies and Reliability of the Data 39
ii. Reliability ..... 39

VI. GENERAL SUMMARY AND CONCLUSIONS ..... 46

VII. Some Applications of the Results of the Present Study to School Uses . . . . . 48

VIII. Some Unsolved Problems Arising from the Present Study . 50

Appendix A . . . . . 55

Appendix B . . . . . 67

# CHAPTER I

## INTRODUCTION

### *I. A Statement of the Problem*

This is a study in curriculum making based upon a sociological determination of one variety of educational values. In one sense it is a study of educational objectives in arithmetic, and an attempt to find out what arithmetic should be taught in the elementary schools of the United States on the basis of what is now actually in use in adult life.

It is proposed to find the optimum amount of arithmetic that actually functions in life aside from vocational uses. An attempt has been made to find not an *absolute* objective in arithmetic but one for the present and the coming generation. We know the minimum requirements in many school subjects but we do not know the optimum requirements. It is the object of our efforts to discover, on the basis of social group usages, the optimum amount, that is, how much and what, non-vocational arithmetic is functioning in our society. The need of such a study is apparent because the offerings in arithmetic in the public schools to-day are largely traditional and vocational. Furthermore, the quality and quantity of the offerings have not been entirely determined by the uses of the adult population. Educational values should be determined by educational needs. There is no known way to discover the actual needs of societies to-day except by means of an objective study and survey of the actual uses of the various elements of our curriculum.

This is essentially a problem in the determination of social values, that is, social, non-vocational uses, in arithmetic. In determining the needs of social groups in arithmetic, or any other subject, we are determining in a sense social values in that subject or field of activity. The objectives in arithmetic which it is the purpose of this study to find are those which may be universally required of all the public elementary school population. The effort has been to separate social, non-vocational or "consumers" uses from vocational requirements.

Before one can thoroughly understand a machine or a process, one must understand its parts and organization, particularly as to function. Similarly, before one can know the value of the different elements of the elementary curriculum one must determine the value of the elements composing the curriculum. Such a breaking up of arithmetical uses enables us to understand the importance of each one. The mistake of mixing vocational with non-vocational objectives will be apparent as we proceed.

Wilson's study of the social and business usages of arithmetic is an example of such a lack of differentiation. He undertook to discover the kind and amounts of business *and* social requirements at one and the same time. To find the business demands, a thorough analysis is necessary. This should be made for each vocation.

After the social, non-vocational arithmetic has been determined it should be an easy next step to analyze specific processes in each vocation. In fact, much good work has already been done in the field of vocational arithmetic, but no extensive investigations have yet been reported where the non-vocational uses have been separated from the vocational.

Vocational arithmetic is important so far as it is required in the various vocations, but in no single vocation, with perhaps a few exceptions, is any great amount necessary. Probably not more than one third of all vocations, such as nursing, sheep-herding, street-cleaning, etc., require any mathematics beyond the four fundamental operations in arithmetic. This would indicate that, after all, vocational mathematics is not so universally important as some would have us believe, although what is required is very necessary and should be thoroughly mastered. If we consider the sum total of arithmetic needs in all walks of life, the amount is prodigious. That the elementary schools should prepare for all vocations is a belief which has often prevailed and one which has apparently directed much of the practice that has resulted in the overburdening of the present curriculum in arithmetic.

### *II. Curriculum Making—a Continuous Process*

An attempt to extend our inquiry to all future generations would partake of the problem of philosophical ultimates. To discover what the next generation is likely to use, however, is one of the duties of educational sociology, and the devising of ways and means of making this selection also belongs in the province of educational sociology.

Arithmetic should be taught but not *all* arithmetic, and we should transmit to the children of to-morrow only that which they are likely to need. In promoting growth and in developing in the young ability to make proper adjustments to the adult group, we shall succeed by using social heritage as a tool. But social heritage not only serves as a tool for that purpose; it meets future developmental requirements of the child. The results of this study may not be viewed in the light of permanently determining what future generations shall require. Such an aspiration would be utopian.

The very nature of a social problem—and all problems are social in a sense—would prevent there ever being an eternal solution of such a series of problems as are found in curriculum revision. Most social problems cannot be solved once for all in the light of all eternity; note, for example, the transportation problem. We are constantly making adjustments to meet the changing needs in this field. Most problems will recur in many forms and the solution will generally require a modification of the *method.*

A problem arises when disharmony arises between two or more dependent variables, and the complexity of a problem is usually in proportion to the amount of disharmony and the number of factors that are involved in the separate but dependent variables. The dependent variables in this particular study are, on the one hand, the static or traditional curriculum requirements in arithmetic, and, on the other hand, a changing dynamic social group development. Accordingly, curriculum making is, and probably always will be, a continuous process. The reason for this is that society is constantly but slowly changing, being the product of an accumulating human culture as conditioned by heredity. Human culture is constantly changing, growing; it takes culture to beget culture and so the eternal cycle goes on. Changing conditions introduce new elements and factors. New elements bring in new relationships and new relationships modify conditions so that new problems are constantly arising and methods will continually have to be devised for solving them.

One who hopes to establish once for all the quantity and quality of the content of the school's offerings is sure to face disappointment. This is so because modern society is dynamic and hence changeful. This change is not so rapid, however, that one generation may not determine and predict within certain limits what optimum amounts of arithmetic, for example, the next generation will use. What educational sociology as a science and other divisions of the science of education may discover is a *method* and a *technique* by which the content of curricula may be revised and reconstructed on the basis of social uses. Social values are the counterpart of social needs, and vice versa social needs produce social values. Values may be translated into social attitudes.

There is developing to-day an amount of culture beyond man's power to control and beyond his ability to comprehend. This fact makes a method and a technique of curriculum selection and revision imperative. A selection of the elements of this culture which are to form the basis of the content of our universal educational requirements in the elementary grades is our most pressing educational problem to-day.

### *III. Two Views in Curriculum Building*

There are two views in the field of curriculum building. The first emphasizes the adult school of thought—the social group as a basis for determining subject matter. The second places emphasis on child life. The first sets out in advance the subject matter for the child to master. The second wishes to keep the on-going process of child life intact. All non-school and school activity should be subjected to that end. The second also holds that the curriculum is to function now or very soon, and unless it is used now or very soon it will be forgotten. Life is a series of adjustment processes, an effort to maintain an ever-increasing relationship of the child to external environment. Both views agree to this statement of the aims set forth in these two expressions, but the second view probably misplaces the significance on the external environment.

### *IV. The Child Develops in Adjusting to Adult Environment*

It is true that the child is ever trying to make adjustments to his environment, but this external environment is largely that of adult life. Adult life is certainly the end or goal toward which child life is tending, the goal toward which adjustments are to be made. Preparation for child life is not the end except as child life is a means to this adult life. Child life leads to adult life. Child life is largely adult life in miniature. There is not a very wide breach between one stage and the other. The child is born into adult life and is trying from the start to make adjustments to it. Child-life activities are a means to that end. Nearly all play life of the child demonstrates this—playing with dolls, playing soldiers, etc., for example. The usual channels through which the child is enabled to make proper adjustments to this adult environment are family and formal group education. But this formal education has to be adjusted to the child's needs and ability to reach this adult state. Adult life is an environment of symbols, largely linguistic, which the child must learn to interpret.

The adult group not only *should* but actually *does* determine the goal. It determines the social values of our arithmetic content as well as the content values of other subjects. The child's nature (that is, the nature of his abilities) should determine, though it has not always been allowed to do so, the methods of presenting the material, the technique of teaching these processes and contents, the placement and arrangement of the material, and the method attendant on the solution of these difficulties. The child's needs can hardly be the basis for selecting the quantity and ultimate kind of the material to be taught, but his needs and nature can select the method and determine the placement of the material of instruction.

### *V. Culture—a Process of Growth and Discard*

It has been said that in the face of our rapid accumulation of culture we cannot predict the future needs of elementary school pupils from the present needs. Those holding to this point of view are partially in error, because no one can deny the continuity of human culture. If the science of anthropology teaches any one thing, it is that culture is continuous and cumulative. That is to say, there are no sudden breaks in human development; it is slowly evolutionary. It is sufficiently gradual to enable any normal adult individual to make all the adjustments necessary when the change comes. Changes in social group mores are generally much slower than those in material culture. This is not equivalent to saying that culture in our various societies is static. It is very dynamic at the present time; but culture does not change over night. It is somewhat gradual over long periods of time; then sometimes come sudden leaps, in material processes—but never in tradition, which is not revolutionary in its changes. This means that each generation will largely be compelled to formulate curricula for the next. Rather it should be said that the coming generation must depend upon the older adult generation for its inspiration and guidance. Age has always been the inspirer of youth and probably will always continue to be.

To insist on the need of continuous activity on the part of the educational sociologist in making a sociological determination of curricula content is not to argue that the tradition which determines the pattern for the immediately coming generation is wrong, but rather that we outgrow from age to age not *all* tradition but some tradition. That which is found to be no longer generally useful at any time should not be perpetuated in the form of an incrusted past. We have numerous examples of the unconscious discarding (on the part of the group) of cultural elements. Examples in material culture are the supplanting of the hand-driven spinning wheel by the one driven by machine power, the one-horse plow by the machine-drawn tractor, etc. With all the complexity of modern culture, life, for the individual, seems to be getting more simple as the group activities are becoming more and more complex, popular belief to the contrary notwithstanding. Thus it happens that there are many subjects the need for some of which for most people is growing less and less as culture complexes develop. This is true of arithmetic at the present time. Some, but not much, arithmetic is needed by a highly cultured people. The very fact that there have existed most highly cultured people with a knowledge of only the very simplest form of mathematical operations, shows that life of a high type can do without much of the complicated mathematics which the school has generally imposed upon the young student. The Greek and Roman systems of mathematics, for example, were extremely simple.

How has it happened that our elementary curricula have become almost anything but "elementary," particularly in arithmetic?

### *VI. Previous Methods of Curriculum Revision*

It appears that most of the subjects now in the curriculum have been urged on the schools by specialists (or special interests) because they liked the subject and were especially interested in it, or needed it in their particular business, rather than because the subject was demanded by the adult group of the common people and later developed by the specialist. At least, this is true in many subjects. Apparently it is true in the case of arithmetic.

Those who have undertaken to determine arithmetic content by studying occupations have started on the wrong lead, for the public elementary schools do not now train and never have trained for any specific vocation and, perhaps, should never do so. Some say, especially business men, that any scheme which would eliminate from the public school all but social values in arithmetic renders the school, in this case, "impractical."

Workers in nearly every field of educational activity have at one time or another and by some sort of method undertaken the task of revising or constructing a curriculum for the elementary grades. They have been influenced by the slogan of our modern pragmatic America, "Let us be practical," which has led many to hunt out the "practical." But the difficulty of adhering to the practical is that many have warped what is meant by practical into signifying what is practical for business purposes, for financial gain, for professional advancement, etc. In this narrow meaning of "practical" some seem to forget that there is such a thing as an idea, a book, or what not, having practical significance if it gives pleasure merely or enriches leisure. It is said that art and music are practical because they administer to the pleasure-giving phases of life. Every one uses art of some kind each day. So much cannot be said for arithmetic. (See Wilson, *A Survey of the Social and Business Usages of Arithmetic.*)

So far as content is concerned, those properly engaged in elementary curriculum construction must have their study, research, and activities focused on non-vocational values. These non-vocational values must be determined by the technique of educational sociology or by educational researches that have a distinctively social point of view, and not by some aspirational assumptions of arbitrary values which often are set up by groups motivated by selfish interests.

Then, too, some psychologists undertake to determine the content very largely on the basis of the nature of the child—his "native" ability, etc. Those who have held, or do now hold, to the idea of formal discipline or transfer of training have let this idea determine the position and importance of certain "hard" subjects in the curriculum. This has doubtless been true in the case of arithmetic. Those who believe that "interest" is inherited and a quality of mind rather than a condition of adjustment of the individual to his work also would tend to place certain material in the curriculum because of its inherent interest to the child.

Psychologists have properly been back of every attempted reform in curriculum construction and in every movement to revise the course of study. We find instances of this in such courses as the Winnetka Plan and the Pueblo Plan. It is admitted that children of, say, ability group level A can learn more content per unit of time than children of group level B of less ability, quality of the material being constant. But curricula are not made especially for the individual child; they are made in terms of children or adult group activities and needs since the child ultimately must make his adjustments to the usages of adult groups. The child does not select. The group of which he is a member selects. We mean here by the term "group" not the whole of society in the abstract but specifically the group level which has the approximate general level of intelligence and social background of the individual in question. In this statement there is no attempt to minimize the importance of the individual as such. He is the most important consideration in any social group. The main efforts of psychologists have been with the individual and much credit is due them, for their work has given us a vast amount of information about the methods and techniques of teaching.

### *VII. What Is Valuable?*

There can be no criticism of the psychologists because they place emphasis on the individual. But granting that the profession now had perfected all the methods of teaching applicable to any subject—arithmetic, for example—we could still ask the question: How much are we to teach by this or that method which the psychologists have perfected? If we have taught perfectly all arithmetic known, then what? The question of how much *is* valuable and *how* valuable would still be a most important question to answer.

# CHAPTER II

## HISTORICAL SKETCH OF INVESTIGATIONS TOUCHING ON THE PRESENT INVESTIGATION

There is an imposing list of investigations in arithmetic which deal with all phases of teaching the material. It is the purpose of this short historical sketch to discuss only those that relate to the content of the arithmetic course of study in the elementary schools. There are several stages, periods, or currents, which, though not clearly defined, may be distinguished, and which relate to the material of instruction. It is true that the time to be devoted to the subject of arithmetic in the schools, the amount of drill, the method of presenting the material, all depend partly upon what the schools and public have considered the content of most worth. For about one hundred years this content was taken largely for granted.

### *Stages of Development*

In the development of the teaching of arithmetic the several stages may be roughly outlined as follows:

I. The stage of Accretion, that is, the period of acquiring the present content, from about 1647 to the decade beginning 1820.

II. The second stage, designated for convenience as the Pedagogical Period, extending to about 1887.

III. The third stage, properly named the Administrators' Opinion Period in determining the content, from about 1887 to about 1900.

IV. The Comparative or "Paper and Paste" Period. This is a period within a period. It is not entirely extinct.

V. The Intelligence Test and Educational Test Period, extending to the present time.

VI. The Business Man's Opinion Period, extending to the present time.

VII. The Sociological Stage or the period emphasizing the social group as a basis of educational values, extending to the present time.

When educators and the public began to be conscious of a burden in the amount of content gradually accumulating in all subjects of the public school curriculum, investigations of one kind and another were started and they probably have not yet reached their peak. These investigations cover a period of more than one hundred years, beginning approximately with the decade 1820-1830 and extending to the present time. To-day there is much interest in the revision of the arithmetic course in the public schools. But until new methods of investigation are perfected arithmetic investigations are at a standstill.

## *I. Stage of Accretion*

Arithmetic was probably taught very early in New England schools because of the commercial activity which early characterized this section of the country. It should be noted that it was the business interest that dictated the arithmetic needed. This was all well enough at that time. But our difficulty now lies in the fact that some of our present educational requirements reach back to this early, ill-defined period. Until fairly recently we have been guilty of requiring every pupil to take the arithmetic taught in those earlier days regardless of his future vocation. The amount grew constantly until such recent studies as those made by Courtis and Jessup—or at least little attempt was made at reduction. Even now the amount is large and the selection of the content in the curriculum is based mainly upon supposed requirements rather than upon objectively determined values and needs.

In early New England, about 1647, when arithmetic was taught in the grammar school it was very rudimentary. Not only was this true, but among the aristocracy of the educated, arithmetic was looked upon as "common," "vile," "mechanic," because it was the accomplishment of clerks, artisans, tradesmen, and others who had no sign of heraldry. (Monroe, W. S. In U. S. Bureau of Education Bulletin, 1919, No. 10, pp. 9-11). What schools there were at that time operated chiefly as grammar schools and were intended for the teaching of writing; little or no mention of arithmetic is made and it was probably not taught in separate school periods outside of writing lessons.

To the south of Pennsylvania the population was scattered over great plantations, and the education of the masses was almost wholly neglected. The rich employed private tutors. Sometimes instruction was given by the minister or by an indentured servant who possessed some education. The aim of this teaching was usually to prepare for college and it did not include instruction in arithmetic. This would seem to indicate, then, that little use was made of arithmetic except in specific vocations, and this arithmetic was only of the simplest kind.

The arithmetic content apparently gradually grew to immense proportions by the simple method of accretion. Bankers, commercial agents, clerks of all kinds, suggesting either directly or indirectly, emphasized their own particular needs. Doubtless school committees were from time to time chosen from these various callings and this elevation and power gave them prestige and influence which they used to increase the arithmetic requirement in the school for which they had assumed responsibility.

### *II. Pedagogical Period*

The second stage, or the Pedagogical Period, extended to about the year 1887 when Francis Walker conducted a campaign in the Boston schools which resulted in an order by the School Committee of that city that certain topics be eliminated from the school requirements in arithmetic. We have named this the Pedagogical Period because it was at this time that such educators as Pestalozzi, Froebel, Herbart had their great influence. "Profound changes were set on foot in our whole educational conception of method and values in education in the decade 1820-1830. Pestalozzi, one of the greatest men in the initial stages of the critical consideration of the content of the school's offerings, made a monumental contribution. His object teaching profoundly affected many subjects then in the public school curriculum and arithmetic was one subject much changed by his theories and practices."

This was the period when the idea of "formal discipline or transfer of training" grew to such proportions, and much of the arithmetic until recently retained in the school curriculum has been justified on the basis of its value as mental discipline. The war on formal discipline is not yet over; there are still many indications of it in the literature of to-day. In this as in all battles there have been extremes.

The Pedagogical Period was a period of "drill" largely for mental discipline and little thought was directed to the question of whether there could be a waste of time even in mind training. But both sides of the controversy on formal discipline have been guilty of exaggeration and confusion of thought. "Absence of any transfer is as false and probably as harmful a notion to use in curriculum construction as its antithesis, complete and magical transfer through the power of the mind. The problem of transfer at the present time is concerned not with the omnipresence of transfer or its utter absence but with such questions as the following: (1) In what types of skill is there enough transfer to use? (2) On what level of mastery of the complicated skill does transfer in useful amounts begin to operate? (3) In what ways can transfer be facilitated and increased? (4) Are transferred skills acquired in one set of data as strong, as lasting, and as easily operated in another set of data? How much is *lost* by transfer? (5) Is the amount of loss of skill from the original situation to the new situation a function of the dissimilarity of the new situation to the original situation?

"The present status of the transfer problem seems to be: Transfer is often small because we fail to realize that the same skills could be used. If we want transfer or the applications of skills acquired in one situation to operate in another situation, we must train in the ability to look for uses of old skills. The old skills will not transfer by themselves except when the similarities are very close." (Buswell and Judd: *Summary of Investigations Relating to Arithmetic,* p. 152.)

From this Summary of the case of formal discipline it should be clear that it would be highly unprofitable to formulate a curriculum in arithmetic on the basis of the "transfer" that could be derived from any specific category or type of problems in elementary school arithmetic.

### *III. Administrators' Opinion Period*

This and the succeeding "periods" of investigations to determine arithmetic content are parallel currents and extend to the present time. But this period is characterized by attempts of school administrators, superintendents of schools, and the like, to select and classify, merely on the basis of their own or others' opinions, the quantity and kind of arithmetic that should be taught. These opinions are frequently formulated not on data collected in a scientific, objective manner but on mere supposition of what would be useful to the pupil in after-school life. Many leading superintendents formed their arithmetic requirements in this way, printed their courses of study, and administered their school systems accordingly. This furnished the material which was the basis for the activity of the fourth stage.

### *IV. Comparative Period*

The twentieth century ushered in a great reform in education in attempts at economy of time in the schools. During the late decades of the nineteenth century there came a general change in the conception of education. Society began to demand a broader type of schooling. The result was a general expansion of the curriculum. This expansion raised numerous questions as to the possibility of making room for new subjects by eliminating waste in the traditional courses of study. Educators began to question every topic included in the entire offering of the curriculum. Reference may be made here to the report of the Committee of Fifteen, which is full of such questioning. At a later date—1904—the question of waste was forcefully presented by McMurry in an address delivered before the National Education Association.

The comparison of courses of study was one of the earliest methods employed in the study of curriculum construction. A paper reporting the results of a study made by this method was published as early as 1900. The statements made in this article show that the reconstruction of the course in arithmetic was being vigorously discussed at that time, although there were few positive studies on which to base the reconstruction. It is also made evident by the results reported in this first reference and in numerous subsequent references that throughout the history of American schools down to the present time there has been a great lack of uniformity in the practice of school systems in different cities. Later studies using this same comparative method have taken up in great detail the practices of a few cities with reference to the distribution of the major topics of arithmetic, or have collected less detailed materials on an extensive scale. All of these studies show the greatest variety of emphasis and some difference even in the major topics taught. (Buswell and Judd: *Investigations Relating to Arithmetic,* pp. 5 and 10.)

## *V. Intelligence Test Period*

The period designated as the Intelligence Test Period is dominated by the psychologists who have proposed to find out the basis for selecting the content of the curriculum by transferring the emphasis to the innate ability of the child. The child's capacity to get arithmetic should be largely the basis for the determination of the content. It was not until 1908, when Stone published his results, that the measurement movement in arithmetic can be said to have gained sanction. (Stone, Cliff W.: *Arithmetical Abilities, and Some Factors Determining Them.*)

In 1909 Courtis published his first articles and from this date on a flood of articles and monographs has appeared. The agitation for the elimination of formal materials, the introduction of measurements, and the demand for reform resulting from measurements have led, since 1910, to various new lines of inquiry which are aimed at the determination of the topics which shall be retained in the arithmetic course and to the discovery of any new material which should be included. We may distinguish these studies which deal with the content of the course as curriculum studies. They differ from measurement investigations in method and degree of applicability to the problems of classroom teaching.

The testing movement has matured rapidly since 1910 and has contributed to the determination of the content of the curriculum through comparisons of different school systems and the resulting establishment of general standards. Again, in 1913, Courtis reported tests which he made of the arithmetical abilities of a number of employees of a commercial establishment in New York City and recommended explicitly that the practices of the schools be governed by such findings. (Buswell and Judd, pp. 5 and 12.) The general conclusion that simplification of arithmetic is desirable is reinforced by studies which attempt to discover the psychological principles which should underlie the organization of the curriculum. (*Ibid.*, p. 16.) There have been many studies in the United States, going back as far as those of Hall, and others in Europe, which have emphasized the number ideas of children. These studies have not only considered the question of what and how much arithmetic should be included but also when formal arithmetic should be introduced in the grades. The course in arithmetic, as we now have it organized, requires, in addition to a large number of experiences, many ideas which are intelligible only when pupils are taught some of the facts of social life. This is particularly true of nearly all forms of arithmetic beyond the four fundamental operations.

### *VI. Business Man's Opinion Period*

Within the last ten years there has developed a type of curriculum investigation which purports to base its conclusions upon what business men *think* should be required in arithmetic. This type of inquiry has come about because many commercial establishments frequently complain that the pupils who come to them for employment from the public schools cannot spell and cannot add and cannot do various things which business demands. Because of this situation a cry arose that we should teach more arithmetic and that arithmetic should be "practical."

Woody found in his investigation, which he reported in the *Third Yearbook* of the Department of Superintendence (pp. 53-54), that business men insist that the schools emphasize the four fundamentals and teach more mental arithmetic. Some emphasize the need of training in interest, discount, and use of the metric system. Each thinks the requirements of his business should be emphasized in the curriculum in the public schools. This very tendency of "special interests" to suggest and sometimes to dictate the content of the public school course furnishes at least one good explanation of the present overburdening of the arithmetic curriculum to-day. Business is now becoming sufficiently specialized to require special training for the positions that are available in each of its important branches, and there should be schools supported by the state to furnish this special training. But the public schools should not be burdened with the responsibility of training for every business. Woody in his investigation also found that calculating machines and other mechanical devices lessen the need of ability to compute on the part of salespeople.

### *VII. Sociological Period*

There is furthermore a decided disposition at the present time to favor increased emphasis on the application of arithmetic to "social" and economic conditions of the day, such as the saving and lending of money, taxation, public expenditure, life insurance, etc. (Buswell and Judd, p. 12.) Wilson's investigation, *Business and Social Usage of Arithmetic,* is perhaps the best study made in this field of sociological investigations. The bibliography at the end of this chapter lists all such studies. Moore, in the chapter on "Arithmetic" in his *Minimum Course of Study,* also by his method of investigation furnishes an illustration of the efforts to base the arithmetic requirements on social and business needs.

The present study is perhaps the only attempt that has yet been made to separate the "business" from the "social" demands and in this respect is a step in advance of any previous study.

### *VIII. Brief Summary and Criticism*

It appears that progress in formulating definite objectives for a course of study in arithmetic is at an *impasse* at the present time. While there is a decided trend toward a sociological determination of arithmetic objectives there is also much confusion. This possibly arises from lack of a clear analysis of the factors in the problem. To date some of those who are inclined to the sociological slant are mixing social and business functions of arithmetic, and as long as this is done there will be confusion. It should be remembered that it was the demand and pressure of "special interests" that produced the enormous accretions of the content material in arithmetic, and unless vocational needs are clearly separated from social uses we will enter into another cycle and have the same problem to meet that has faced the schools for the past one hundred years. It is true that we cannot separate arithmetic, or anything else in real life, into compartments, but it is absolutely necessary for purposes of analysis and understanding to separate vocational mathematics from non-vocational "social" uses if we are to find our way out of the present maze of curriculum study. Lack of realizable objectives or a proper method of procedure caused many of the investigators of twenty-five or thirty years ago to go from one extreme to another or to fall back on some rationalizing process such as "formal discipline." Many of them have made valuable contributions to the cause of education, but have brought us no nearer to the defining of objectives. This will be demonstrated by consulting the list of textbooks classified according to time allowed the various categories commonly included in texts for the schools. This table, given by Buswell and Judd (p. 29), shows that textbook writers even up to the present time have shown just as much uncertainty about what to include or to exclude and how much time to require as has been shown by administrators and teachers.

1. The process of elimination and refinement in the arithmetic content has been going on in one way or another since the beginning of the nineteenth century. Warren Colburn led the way in trying to answer the question in "What is Essential?", when in 1819 he showed that time was being wasted and something should be done about it. President Francis A. Walker in 1887 continued to stress the importance of economy of time in the grades, and President Harper in 1902 and Dr. Frank M. McMurry more recently have continued the attack.

2. The profession began to cast about for a remedy. Something had to be eliminated. Suggestions were made that certain topics be omitted but no one had proof of what was of most value and what of least value.

3. The less original administrators began to copy courses of study made by more progressive school superintendents. But even with this type of elimination as an example, school authorities and teachers, being in general a conservative group, have been slow to discard anything in the curriculum that has been long an established part of the offerings of the school.

4. The National Education Association in various committees appointed from time to time have made recommendations, but these have not been based upon any well-defined objective nor have methods been perfected that have thus far brought results.

5. The formal discipline belief having been largely discarded, the next step was to turn to the business interests for opinions on what should be included. In the light of the present investigation this is certainly not a correct way to determine elementary arithmetic content. It seems to result in as much confusion of thought as we had when we relied upon the opinions of school administrators. It fosters the fallacious notion that all boys and girls who go through the elementary schools should be equipped to use arithmetic required in any business. Probably most people would deny this. But if we accede to the proposition that any one business should demand that its needs be included, then all businesses, since they are taxed to support the public schools, have the same right to dictate. The consequence of this policy cannot be overlooked.

6. Finally we have those who have attempted to determine the business and social needs of arithmetic on the basis of what is actually used in business and society. Wilson's study is an illustration of this movement; Moore's investigation is another. In these studies there are mixed aims, business needs on the one hand and society's needs aside from business on the other. They have failed to analyze the problem into its elements and are probably on the wrong lead.

7. The present study avoids this difficulty by confining its efforts to the determination of "consumers," that is, social needs aside from vocational requirements. This done, the vocational or business needs will be relatively easy to determine.

## *IX. Bibliography*

[Contains only references that relate to investigations concerning Arithmetic content.]

1. BROWN, JOSEPH C. *Curricula in Mathematics.* U. S. Bureau of Education Bulletin No. 45, 1914.
2. BURNHAM, WILLIAM H. "Arithmetic and School Hygiene." *Pedagogical Seminary* XVIII (March, 1911), pp. 54-73.
3. CAMERER, ALICE. "What Should Be the Minimal Information about Banking?" Third Report of the Committee on Economy of Time in Education, pp. 18-26 in the *Seventeenth Yearbook* of the National Society for the Study of Education, Part I. Bloomington, Ill., Public School Publishing Company, 1918.
4. CHASE, SARA E. "Waste in Arithmetic." *Teachers College Record,* XVIII (September, 1917), pp. 360-70.
5. COURTIS, S. A. "The Courtis Tests in Arithmetic." *Report on Educational Aspects of the Public School System of the City of New York to the Committee on School Inquiry of the Board of Estimates and Apportionment,* Vol. I, Part II, Subdivision I, Section D, pp. 391-546. City of New York, 1913.
6. GLASS, JAMES M. *Curriculum Practices in the Junior High School and Grades 5 and 6.* Supplementary Educational Monographs, No. 25. Chicago; Department of Education, University of Chicago, 1924, p. 182.
7. GREENWOOD, J. M. "The Evolution of Arithmetic in the United States." *Education,* XX (December, 1899), pp. 193-201.
8. GUILER, WALTER S. "Curriculum Building in Arithmetic." *Educational Research Bulletin,* III (Feb. 6, 1924), pp. 49-54. Columbus, O., Bureau of Educational Research, Ohio State University.
9. HARRIS, ALICE L. "Number." *A Kindergarten-First Grade Curriculum,* pp. 52-56. U. S. Bureau of Education Bulletin No. 15, 1922.
10. JACKSON, LAMBERT L. *The Educational Significance of Sixteenth Century Arithmetic from the Point of View of the Present Time.* Teachers

College Contributions to Education, No. 8. Teachers College, Columbia University, 1906.

11. Jessup, Walter A. "Some Eliminations in the Content of Arithmetic as a Factor in the Economy of Time." *Proceedings and Addresses of the N. E. A.,* LI (1913), pp. 464-68.

12. Jessup, Walter A. "Economy of Time in Arithmetic." *Proceedings and Addresses of the N. E. A.,* LII (1914), pp. 209-22.

13. Jessup, Walter A. "Economy of Time in Arithmetic." *Elementary School Teacher,* XIV (June, 1914), pp. 461-76.

14. Jessup, Walter A. "Current Practices and Standards in Arithmetic." *Minimum Essentials in Elementary School Subjects—Standards and Current Practices,* pp. 116-30. *Fourteenth Yearbook* of the National Society for the Study of Education, Part I. Chicago, University of Chicago Press, 1915.

15. Monroe, Walter S. *Development of Arithmetic as a School Subject.* U. S. Bureau of Education Bulletin No. 10, 1917, p. 170.

16. Moore, Ernest C. *Minimum Course of Study,* pp. 1-41. New York, Macmillan Co., 1922.

17. Smith, David Eugene. *The Progress of Arithmetic in the Last Quarter of a Century.* Boston, Ginn and Co., 1923.

18. Smith, Nila B. "An Investigation of the Uses of Arithmetic in the Out-of-School Life of First Grade Children." *Elementary School Journal,* XXIV (April, 1924), pp. 621-26.

19. Spaulding, F. T. "An Analysis of the Content of Six Third-Grade Arithmetics." *Journal of Educational Research,* IV (December, 1921), pp. 412-23.

20. Starch, Daniel. "Transfer of Training in Arithmetic Operations." *Journal of Educational Psychology* (June, 1911), pp. 306-10.

21. Stone, Cliff W. "Arithmetic Considered as a Utilitarian Study—What Should Be the Course of Study?" *Elementary School Teacher,* III (April, 1903), pp. 533-42.

22. Taylor, Joseph S. "Omitting Arithmetic in the First Year." *Educational Administration and Supervision,* II (February, 1916), pp. 87-93.

23. Wilson, G. M. "A Survey for the Social and Business Use of Arithmetic." *Second Report of the Committee on Minimum Essentials in Elementary School Subjects,* pp. 128-42. *Sixteenth Yearbook* of the National Society for the Study of Education, Part I. Bloomington, Ill., 1917.

24. Wilson, G. M. *A Survey of the Social and Business Usage of Arithmetic.* Teachers College Contributions to Education, No. 100. Teachers College, Columbia University, 1919.

25. Wise, Carl T. "A Survey of Arithmetical Problems Arising in Various Occupations." *Elementary School Journal,* XX (October, 1919), pp. 118-36.

26. Bobbitt, Franklin. "Technique of Curriculum-Making in Arithmetic." *Elementary School Journal,* XXV (October, 1924), pp. 127-43.

27. Buswell and Judd. *Investigations Relating to Arithmetic.* Supplementary Educational Monographs, No. 27, Chicago; Department of Edu-

cation, University of Chicago. (Has a most extensive and complete Bibliography with annotations.)

28. BOBBITT, FRANKLIN. *How to Make a Curriculum.* Boston, Houghton Mifflin Co.
29. CHARTERS, W. W. *Curriculum Construction.* Chicago, University of Chicago Press.
30. CALLIXTA, SISTER M. "The Educational Objectives of Arithmetic." Catholic University of America, *Educational Research Bulletins,* Vol. I, No. 3, p. 28. Washington, Catholic Education Press, 1925.
31. STITT, EDWARD W. "School and Business Arithmetic—Limitations and Improvements." *Proceedings and Addresses of the N. E. A.,* XXXVIII (1900), pp. 566-72.

# CHAPTER III

## METHOD USED IN THE PRESENT STUDY

### *I. Method in General*

The method employed in this study is one suited to the investigation and measurement of social values. It is based upon a general hypothesis of the validity of estimates of competent persons in determining qualitatively and quantitatively social values, social needs, and social interactions. Most of the objects of measurement to be mentioned here cannot be measured by the same type of measuring unit and with the objectivity in exactness so prevalent in the physical sciences.

It has not yet been demonstrated that such types of measurement and exactness as are used in the physical sciences are altogether desirable when applied to the estimation of social values—even if it were possible so to apply them. What those who are dealing in social values most need and desire to know is the *trend* of social interactions and reactions. This objective of establishing trends and central tendencies is also one of the ideals of statistical procedure, that is, to establish limits within which the central tendencies may be found.

The tools of a sociologist and his techniques are elusive and difficult of comprehension—at least some of them are—to all but those who are trained in the interpretation and use of these techniques. No technical manipulation of formulæ can be understood by the great mass of people, nor completely comprehended by scientists themselves unless they are trained in dealing with social data and in its interpretation.

Probably it is to personal judgments or estimates that objection may be raised. Under certain conditions opinions are facts, for we must remember that data of any kind are not facts in themselves, and do not become facts until critically analyzed; hence facts are products of the human mind. They do not exist free in nature. Opinions of those competent to express them, when they are collected and given critical analysis and related to social interaction processes, become social facts and are socially useful. (Park, Robert E. In *Publications of the American Sociological Society,* Vol. XVIII, p. 105.)

Subjective judgments as to the social value of certain types of information, when gleaned from competent persons, have value. One great defect, however, in such a collection of judgments in the past and even now is the method used.

Pooling subjective judgments of competent individuals on social values for the next twenty or thirty years will give fairly reliable results in determining educational values; and objectivity in many subjects is not generally possible now and probably will not be possible for some time to come.

### *II. Studies Based Upon Opinion and Estimate*

A. It is maintained here that many of our most valuable studies in the realm of curriculum construction have been based upon opinion or estimate. We mean by "estimate" a judgment or opinion usually implying careful consideration or research, and careful weighing of observational data. For example, the handwriting scales, Thorndike's and others, have in them elements of comparative judgments or estimates. How else could the best quality of penmanship be determined than by judgments of competent persons? Of course legibility can in some measure be subjected to the time unit of measurement of production and easy interpretation. These values, however, are based upon norms of large groups.

B. Ayres' spelling list and those of others are similarly social data. True, these men had before them objective material in the form of pupils' composition, etc., with the misspelled words on them. Jones's "Hundred Demons" is an example of this type of word study.

C. Recently Dr. Otto Hamilton presented a dissertation at Teachers College, data for which were obtained from questionnaires and reports from teacher-training institutions in all sections of the United States. Aside from statistics on such matters as money actually spent and pupils actually enrolled in school, etc., much of his data are based upon some one's estimate.

D. Wilson's studies, one of which served as a Doctor's dissertation at Teachers College, are based upon estimates. From the point of view of the present study these methods are proper when due impartiality of attitude is maintained.

E. Much historical data furnished by our best historians have been gleaned from original material, from old manuscripts often written by a biased eye-witness or by persons emotionally urged, religious advocates, or political writers of bias. And yet reliance and faith are placed in these documents. On these sources are based scientific histories of past ages. One is justified in saying that no one is impartial, even military experts, in recording the total happenings in a battle or campaign. No writer of the Civil War period was unbiased in his selection of data which he included in his writings concerning the causes or results of the Civil War. No better proof of this can be offered than to cite the number of histories partial to the North or to the South. No one can free himself from bias. We may be justified in saying that an eye-witness of the Battle of Hastings in 1066 A.D. had lapses of memory and did not note all the happenings of importance as they actually occurred. There is no condemnation of the student of history for indulging in this sort of investigation. The fact is that it is almost his only means of getting at a social situation long past.

Even if the judgments of those recording events are partly wrong or partly biased, it does not always greatly matter since, in reality, it is the effect on him—the writer—and others who lived at the time that is most important. And these things that are really essential socially are social facts not to be overlooked. They are measured by their effect and the social *trend* they set up. The elements that control count.

### *III. Remembered Use—the Basis of This Study*

The present study is not based upon judgments of persons as mere judgments but rather upon actual use or remembered use of certain types or categories in arithmetic, such as Compound Denominate Numbers or Factoring. Certain groups were asked to check the kinds of arithmetic used in their daily lives (see Appendices A and B). An objection may be raised that in checking the kind of arithmetic he uses one would not remember all he has used. To this it is agreed. So do all the historical writers commit to writing only what they remember. This forgetting may be a basis for objection to the present method of determining arithmetical usage. It is not a valid objection, however. If those checking the arithmetical problems which they have used, or are now using, should not remember every calculation that they have made, what they do *not* remember would not matter materially since what they have used only a few times and have forgotten has not affected their lives to any great extent and could not, therefore, be an important part of their training or needed equipment in arithmetic. It is this very problem of finding the *important kind* of arithmetic that this investigation proposes to solve.

Again, it is hardly to be supposed that the six hundred and eighty-two persons checking the blanks used in this study would have forgotten approximately the same categories or the same problems as the investigation here demonstrates. It is significant, indeed, that most of them should have used and forgotten the same thing. That fact alone should answer the objection to the method used.

In the field of physical science much of the "accurate" data are approximations. All that science can ever do is to point out probabilities, to indicate trends. If competent estimates can give actual trends of social data, even approximations, they may be so organized as to be classed as scientific, and thus be highly useful to society. Nor can science measure now, in many cases, quality or quantity by any other measure than critical estimates. Many of the tests constructed to measure intelligence are dependent to some extent upon interpretations and judgments. In this way, then, there are not yet many objective measuring units for anything except physical science. Much of the measurement in social science is in terms of descriptive words, as "very good," "better," "very bad," and the like.

Since social science has at the present time few exact measurements for social data, a careful definition of terms is necessary, and a keen, incisive analysis of social situations and interactions should be made. In one sense of the word all objective measurement is a form of analysis and definition—delimitation. Measurement is analysis. The object of measurement is definition, finding limits, bounds. In dealing with many social phenomena, attempts at objective measurement, for the present at least, must be partially discarded and a resort to strictly clear analysis used, chiefly because most social phenomena cannot be separated entirely from opinion and bias.

Astronomical science has long been in possession of an accurate method of calculating exactly, even in terms of seconds, when and where a total eclipse of the sun or moon will occur. So thoroughly have astronomers worked out this calculation that they can predict accurately years ahead just when these events will transpire. But most phases of social science are not so fortunate. When this complicated science of humanity develops its methods to a point where the social need and therefore social values can be determined—the method of finding out what and how much of human culture will be needed in the universal curricula for all normal persons—then the matter of building the curriculum will be much easier and more definite than it is at the present time. But even then, the solution cannot be a universal formula such as we now use in working out an accurate prediction of the eclipse of the sun, for we are dealing with a dynamic and changing factor and one consequently subject to change from one generation to another. These changes, however, may be roughly predicted.

### *IV. Method Used in Solution of This Problem*

The method used in the solution of this problem of finding the amount and kind of "consumers" arithmetic is not based upon opinion but upon remembered use of certain type problems which are listed in Appendices A and B. No one was asked to state his opinion. The writer prefers to designate the blanks used in this study as "answeraire" blanks, which are so arranged that the answers are checked with great facility and accuracy, no written statements or calculations being required.

The same criticism and requirement for scientific accuracy hold for this "answeraire" as are held valid for the usual questionnaire. The value of a recollection questionnaire study depends upon: (1) the nature of the topic to be investigated; (2) the care and skill with which the questions are formulated; (3) the composition and representative character of the group of persons answering; (4) the honesty and conscientiousness of those who answer; (5) the scientific ability of the investigator who attempts to interpret the data or to draw deductions from it; (6) the kind of use that is to be made of such deductions.

## *V. Analysis and Description of Method Followed*

A. The writer collected copies of all arithmetics now on the market and those that have been in use within the last quarter of a century. These texts were examined to determine the types of arithmetic problems that have made up the subject matter of arithmetic instruction in the public schools of the United States. From this search a list of problems was selected which included every category or type of problem printed in the texts examined. By the term "category" is meant a heading for classifying problems, such as Proportion or Square Root. By the term "type" is meant a problem which represents the principles involved under each category. The main categories are as follows:

1. Compound Denominate Numbers.
   Sub-categories:
   *a.* United States Money.
   *b.* Long Measure.
   *c.* Square Measure.
   *d.* Dry Measure.
   *e.* Liquid Measure.
   *f.* Cubic Measure.
   *g.* Metric System.
   *h.* Apothecary's Measure.
   *i.* Other Measures.
2. Factoring, simple and compound.
3. Prime Factors and Complete Numbers.
4. Fractions, including simple, complex, and decimal.
5. Interest, including simple, compound, and annual.
6. Discount, including bank discount, true discount.
7. Percentage, including the five cases.
8. Aliquot Parts of 100, such as 12½.
9. Mensuration.
10. Involution.
11. Evolution.
12. Roots, including square and cube root.
13. Proportion, including simple and compound.
14. Ratio, including simple and compound.
15. Progression, including arithmetical and geometrical.
16. Exchange, both foreign and domestic.
17. Insurance, both life and fire.
18. Taxes and Duties.

The four fundamental operations of arithmetic, addition, subtraction, multiplication, and division of simple whole numbers, were assumed as universally used in some form, in some degree at least, by every normal literate adult and hence are not included in this study.

B. Under the above classifications 345 problems were selected (see Appendix A). These were arranged in the following manner: The "answeraire" was prepared in such a way that the one who was asked to check in appropriate spaces the degree of use could do so without having to write any statements or make any calculations. Three type problems under each category were selected, one very simple, one of medium difficulty, and one representing the greatest difficulty found in the texts. The basis for determining the "difficulty" of the type problem was the complexity and number of operations necessary to the solution of the particular problem.

C. The next step was to analyze all the life situations in which arithmetic could be used aside from vocational need. By "life situation" is meant any situation in which a need of any kind of knowledge of arithmetic could arise.

In this study only those life situations which have to do with the use of arithmetic aside from one's vocation have been used. These are:

1. Buying at the store for home consumption.
2. Making change.
3. Reading newspapers, periodicals, and magazines.
4. Investing one's savings, aside from one's own business investments.
5. Writing letters, etc.
6. Traveling, consulting time tables, etc.
7. Recreation, golf, bridge, billiards, etc.

Doubtless there are many values in a knowledge of arithmetic that are not exclusively non-vocational in character. Among these may be mentioned:

*a.* Number sense. The "feeling" for mathematical relationships —a sense of quantity, proportion, etc.

*b.* Mathematical appreciation, comprehension. This probably may be classed as an example of the "number sense."

*c.* Love of mathematics. There is a bare possibility that a few rare persons so mathematically inclined get much enjoyment from solving very intricate and difficult arithmetical problems involving difficult operations in mensuration, roots, solution, compound fractions, proportion, etc. Such individuals are so few in number that they need not be considered to any appreciable degree in our public school curriculum.

*d.* According to some, one of the values in the study of arithmetic is mind training or the "transfer of training." This is a moot question and no one knows enough about it to be dogmatic. (Buswell and Judd, pp. 150-52.)

But values listed above under *a, b, c, d* are "derived" or "secondary" or "by-product" values as compared with those listed under 1, 2, 3, 4, 5, 6. So far as we know, these by-product values are not used in the direct sense in which one uses arithmetic in, say, buying a peck of potatoes.

These by-product values are the result of the *method* of teaching and presentation of the content which we have found to be useful. We should say, rather, that the values mentioned in 1 to 6 are the result of specific categories and the by-product values result from all the categories combined. Secondary values may be realized by studying any content; they are not the product of any particular content. (Buswell and Judd, Chaps. 7, 8, 9, 10; Freeman, *Psychology of the Common Branches,* Chap. IX, p. 179-237 and references.)

There are some non-vocational life situations which require no arithmetic beyond the four fundamental operations of addition, subtraction, multiplication, and division of simple whole numbers. *Recreational mathematics* we will call it—bridge, golf and other forms of athletics and social games. An analysis of these forms of recreational arithmetic shows the simple form of arithmetic required. Certainly no form of arithmetic goes much, if at all, beyond operation of simple whole numbers.

In the examination of the various texts, it was found that Ray's *Practical Arithmetic* and similar texts used twenty to thirty years ago contained all the categories and types of problems that are contained in texts of later date. The array of types or topics in such books represents the kind and amount of arithmetic taught from two to five decades ago to boys and girls, now adults ranging in age from 35 to 60 years.

Our problem is in reality to find out just what particular kinds of this arithmetic represented by the above types of problems and given in Appendix A are actually functioning in life to-day, barring vocational usage. Much more arithmetic was taught in the public schools thirty-five years ago than is taught at the present time. This will be amply demonstrated when one examines the most popular arithmetics now in use and compares their content with the content of texts of a generation ago. Then children were taught a prodigious amount of arithmetic, and it is an interesting problem to learn how much they now use.

D. Reference to Appendix A will illustrate the method of arranging the 345 type problems. It will also show that under each of the main headings six degrees of frequency of use are indicated with appropriate spaces for checking. They are: 1. Very often; 2. Often; 3. Occasionally; 4. Seldom; 5. Very seldom; 6. Never.

The "answeraires" were administered by the writer and by two persons who were thoroughly familiar with the purpose and method of the study. The blanks were distributed to individuals in the following places:

| City | Population | No. of Returns |
|---|---|---|
| Silver City, N. M. .. | 3000 | 18 |
| Glen Ellyn, Ill. ..... | 2851 | 12 |
| Richmond, Ky. ..... | 5622 | 12 |
| Total ........... | ............... | 42 |

Since the blanks which were to be filled out by checking contained thirty-nine pages, considerable time and pains were used in helping those who filled them out. Those asked to fill out the blanks were chosen, as nearly as possible, from all walks of life. The blanks were collected, the data tabulated and weighted, and the results compared.

Careful scrutiny of the data furnished by blank A will show that, barring certain problems, the use of these type problems was so slight that most of them could be eliminated from further consideration. There was a scattering number of problems that showed considerable variation in use. These were selected as a basis for the content of another "answeraire" (Appendix B). Care was exercised to see that at least one problem from each category and type in the first blank A was included in blank B.

This second blank was constructed, as will be seen by reference to Appendix B, so that those asked to check it could answer by checking under "yes" or "no" under the six life situations used in the first blank, with the addition of one "Filling Out Income Tax Sheets." Copies of blank B were sent to city school superintendents in the following cities with the instruction that they be given to the sixth grade teachers to hand out to each sixth grade pupil. These pupils carried the blanks to their parents or guardians to be filled in and returned to the superintendents as quickly as possible. The superintendents returned the blanks to the writer and the results were tabulated and correlations of non-use in each of the seven life situations were obtained.

The cities are as follows:

| City | Population | No. of Complete Returns |
|---|---|---|
| Norwood, Mass. .... | 12,627 | 49 |
| Bloomfield, N. J. ... | 22,011 | 140 |
| Tupelo, Miss. ...... | 5,055 | 62 |
| Oregon City, Ore. .. | 5,686 | 30 |
| Portsmouth, Ohio .. | 33,011 | 217 |
| El Paso, Texas ..... | 106,000 | 142 |

## *VI. Method Used in This Investigation Compared with Methods Used in Previous Investigations*

The method of investigation described above differs in some respects from that of any previous study designed to determine the amount and kind of arithmetic used by the adult population of the United States.

Miss Alice Camerer made an investigation to find out what should be the minimum information about banking.[1] In this study, information about certain businesses and their processes was found to be as important as ability to make arithmetical computation. A list of fifty-five items was sent to a number of banking employees and parents to be ranked according to importance in

[1] Published in the *Seventeenth Yearbook,* National Society for the Study of Education, Part I. pp. 19-26.

the general knowledge of the people of the community. Though this is a valuable study, it fails to make any distinction between "consumers" or non-vocational arithmetic and vocational uses. Knowledge of arithmetic used in a bank is, indeed, needful, but it seems to the writer that to use this method to determine the amount of arithmetic to be included in the curriculum of the elementary schools is a confusion of aims. It is such confusion or lack of analysis of the two uses of arithmetic—vocational and non-vocational—that has probably prevented educators from coming to any agreement on "minimum" essentials in the elementary school course. Even textbook writers are confused. Buswell and Judd in their *Investigations Relating to Arithmetic* (pp. 26-36) show that some of our most modern arithmetics lay more emphasis on some categories in arithmetic which have long been thought to be of little value than on others which are more useful. In fact, there is no uniformity in the amount of space and time required for the various categories in texts.

It should not be forgotten that the public elementary schools are not now nor have they ever been vocational schools. From their very nature they can never function as such. It is true that a study like Miss Camerer's is valuable in furnishing items of importance in the subject of "informational" arithmetic, but the erroneous idea may develop that enough of this sort of subject matter "about" arithmetic could be taught in the public schools to prevent the adult population of the coming generation from falling victims to "shysters." Against this it should be urged that it is impossible to teach much of this sort of thing in the public schools. Pupils should be taught to seek advice from experts on banking subjects. They should be taught to rely upon the legal or business expert for such information and guidance. The public schools cannot develop experts. It should be pointed out that in Miss Camerer's study the attempt is made to determine the amount of use of such arithmetic on the basis of the opinion of business experts rather than the actual amount used by the common man, not a banker, in his banking operations.

Mr. Elmer Hanson made a study of "The Arithmetic of Sales Persons' Tally Cards" in 1924 and the results are published in the *Third Yearbook* of the Department of Superintendence (p. 56). This investigation is based upon business men's opinions of the uses of arithmetic. The results show that these business men do not know what is needed except in their own business. Their opinions are no better than the opinion of any one else, when it comes to determining what arithmetic the common man needs or actually uses in everyday life. These men listed what is used in each vocation, but this study fails to separate the needs of business vocations from actual non-vocational uses. The topics which they recommend are:

1. Building and Loan Associations.
2. Keeping Simple Accounts.
3. Investing Money.
4. Bonds as Investments.
5. Real Estate as Investment (cheap rentals, good residence property, business blocks, or farm lands as investments).
6. Marks of a Good Investment (it is estimated that the American people are fleeced out of $60,000,000 a year on get-rich-quick concerns).
7. Taxes, Levies, Public Expenditures, etc.
8. Profits in Different Lines of Business.
9. Life Insurance as Protection and Investment.

This sort of study gets nowhere in the matter of curriculum construction, for it fails to analyze its elements. It is true that the information given is valuable, but how valuable and how much is advisable—a question that the curriculum maker must answer—is not indicated by such studies as these.

The public schools, in reality, have never been equipped to give information on all the subjects that business men and other groups would insert in the public school curriculum. That is the province of vocational schools. Such attempts on the part of business men to dictate the curriculum content have made our offerings burdensome and in some cases impossible to-day. Instead of trying to include all this information the public school should try to teach its pupils to follow experts and not to risk making technical decisions for themselves.

The investigation of Clifford Woody on the types of arithmetic used in certain types of salesmanship, published in the *Elementary School Journal* (Vol. 22, No. 7, March 1922, pp. 505-521) is a valuable contribution, but it also fails to meet the point of the present investigation in that it is not selective as to the kind of arithmetic used by the common adult population. It also mixes vocational and non-vocational aims. The problem on this basis is impossible of solution. Woody found that adult needs in percentage were as follows:

1. 20 per cent used percentage 31 times.
2. 10 per cent used percentage 26 times.
3. 25 per cent used percentage 3 times.
4. 50 per cent used percentage 13 times.

He also found that the percentage used was of the simplest kind and appeared in simple accounts, and that very little was used at that. He found that percentage was used only in respect to United States money. These conclusions are also reached by Calloway in a study of "Mathematics Needed in a Clothing Course for Freshmen." Professor W. W. Charters also, in his *Curriculum Construction* (p. 23-36), found little use of decimals.

Wilson's study has already been mentioned as to method. His is one of the most valuable recent studies that have been made. But he, too, has mixed consumers with vocational uses and thus has confused rather than solved the problem—"How much and what arithmetic shall we teach?" Charters, Wilson, Moore, and others in their investigations to determine the answer to this question have used the method of "estimates" and the method of "opinion." They have all mixed the aims involved in the issue here under consideration.

# CHAPTER IV

## PRESENTATION OF THE DATA

The cities used as bases for collecting data and the population of each were given in Chapter III. The present chapter will be occupied with the presentation of the data in analytical form conforming as nearly as possible to Appendices A and B.

The data submitted here come from two "answeraires" represented by Appendices A and B. Appendix A blank was administered by the author and two assistants in three cities. Forty-two blanks were completely and carefully filled out. As has already been stated, this blank contained 39 pages and 345 problems. When the data resulting from this "answeraire" were tabulated and weighted on a percentage basis, it was clear that a number of the problems were used very little or were not used at all. These problems were at once eliminated on the basis that problems which showed less than 25 per cent use by those filling out the blank could be discarded. There were certain problems on which there was not close agreement about use because of too few subjects. These and the simplest problem under each type and category were selected to form the basis for the second blank, Appendix B.

On the first page of each blank were questions concerning age, schooling, frequency of checking bank statements and grocery bills, etc., newspapers read; and whether the one questioned supported a family. Results from these answers will be presented in the next chapter on central tendencies of the data.

When all the blanks were returned, there were several from each city which, for one reason or another, were not complete. Some were not filled out in every detail; others indicated that those filling them out failed to understand just what was desired; and still others indicated by the manner in which the answers were checked that the "answeraire" was not taken seriously.

All such blanks were discarded. With these eliminations there remained 682 blanks properly filled out, and these are considered reliable from the standpoint of this study.

It will be noticed that the first blank contained six degrees of frequency of use. When the per cents were worked out for each degree of frequency, 70 problems showed a use of "Very Often." These were used by less than 15 per cent of those filling out the blank. The rest were checked to indicate use from "Occasionally" to "Never."

It was decided, therefore, to add all the per cents of use or non-use of each problem and to list them under the two "Yes" and "No" columns, which was the method of checking in the second blank. This made the two blanks more comparable. Examination of the data in this chapter will make this clear. It will show that there is rather close correlation between the use and non-use of the same problems in each blank. Since the first blank was primarily for the purpose of selecting the problems for a second study or blank, correlations were not worked out. The main body of the dissertation is based upon the second blank and conclusions are largely formulated on the data furnished by this blank.*

Chapter V, which shows central tendencies in the data, indicates the general trend of the use of the problems listed in the Appendices.

* The data obtained from the results of the blanks (Appendix A and Appendix B) on use and non-use are not printed here, but are included in the manuscript copies of this dissertation on file in the Library of Teachers College, Columbia University.

# CHAPTER V

## CENTRAL TENDENCIES IN THE DATA

The first sheet in both blanks represented by Appendices A and B gives data on (1) ages of those filling out the blanks; (2) schooling; (3) whether the one questioned supported a family; (4) whether he checked bills from the store and bank statements; (5) sex; (6) the newspapers and magazines read; (7) occupation followed.

Tables I and II give data on items (1) to (5) for those filling out the blanks in the two groups of cities.

TABLE I

| *City* | *Bloomfield* | *El Paso* | *Norwood* | *Oregon City* | *Portsmouth* | *Tupelo* |
|---|---|---|---|---|---|---|
| | | | AGE | | | |
| Range | 18–57 | 25–65 | 24–72 | 28–68 | 28–68 | 30–56 |
| Median | 36.4 | 39.3 | 43.3 | 43.3 | 40.2 | 38.8 |
| Q1 | 33.22 | 36.69 | 34.0 | 36.81 | 34.6 | 34.2 |
| Q3 | 42.25 | 42.83 | 51.0 | 52.25 | 43.7 | 46.1 |
| | | | SCHOOLING | | | |
| Range | 1–4 coll. | 1–4 | 1–4 | 5th–2 coll. | 1–4 coll. | 3–4 coll |
| Median | 8.5 | 10th | 8th | 8th | 8th | 9th |
| Q1 | 7.4 | 8th | 7th | 7th | 6th | 8th |
| Q3 | 9.6 | 12th | 9th | 9th | 9th | 10th |
| | | | SUPPORT A FAMILY | | | |
| % Yes | 69 | 77 | 95 | 91 | 70 | 79 |
| % No | 31 | 23 | 9 | 9 | 30 | 27 |
| | | | (a) CHECK BILLS FROM STORE | | | |
| % Yes | 95 | 100 | 95 | 100 | 98 | 100 |
| % No | 5 | 0 | 5 | 0 | 2 | 0 |
| | | | (b) CHECK BANK STATEMENTS | | | |
| % Yes | 87 | 91 | 76 | 91 | 77 | 83 |
| % No | 13 | 9 | 24 | 9 | 23 | 17 |
| | | | SEX | | | |
| % Male | 54 | 62 | 70 | 63 | 53 | 54 |
| % Female | 46 | 38 | 30 | 37 | 47 | 46 |

TABLE II

| | *Glenn Ellyn* | *Richmond* | *Silver City* |
|---|---|---|---|
| AGE | | | |
| Range | 27–56 | 24–57 | 24–57 |
| Median | 46 | 34 | 43 |
| Q1 | 39 | 39 | 36 |
| Q3 | 55 | 42 | 52 |
| SCHOOLING | | | |
| Range | 9th–coll. 4 | 6th–coll. 4 | 6th–coll. 4 |
| Median | coll. 1 | 12th grade | 11th grade |
| Q1 | 10th grade | 9th grade | 8th grade |
| Q3 | coll. 2 | coll. 2 | coll. 2 |
| SUPPORT A FAMILY | | | |
| % Yes | 84 | 75 | 75 |
| % No | 16 | 25 | 25 |
| CHECK BILLS | | | |
| % Yes | 100 | 90 | 90 |
| % No | 0 | 10 | 10 |
| CHECK BANK STATEMENTS | | | |
| % Yes | 90 | 90 | 90 |
| % No | 10 | 10 | 10 |

Table III shows in per cents the non-use of the various type problems represented by Appendix B. All distributions, step intervals, averages, etc., are based upon per cents of non-use in each life situation.

TABLE III—NON-USE

I. BUYING AT THE STORE, ETC.

(Appendix B Blank) $P.E._{av.} = \frac{.6745\sigma \text{ (dis.)}}{\sqrt{123}}$

| *Cities* | *Range* | *Median* | *Q1* | *Q3* | *Av.* | *Sigma* | *P. E. (Av.)* |
|---|---|---|---|---|---|---|---|
| Bloomfield | 28–100 | 88.59 | 75.98 | 93.55 | 83.45 | 11.95 | .726 |
| El Paso | 35–100 | 82.50 | 71.75 | 88.71 | 77.70 | 13.10 | .796 |
| Norwood | 65–100 | 95.29 | 75.50 | 91.89 | 92.33 | 13.05 | .793 |
| Oregon City | 45–100 | 86.71 | 76.15 | 92.53 | 82.5 | 17.85 | 1.026 |
| Portsmouth | 45–100 | 86.71 | 76.15 | 92.53 | 82.58 | 13.60 | .826 |
| Tupelo | 35–100 | 72.78 | 62.34 | 80.83 | 68.00 | 12.15 | .738 |
| II. MAKING CHANGE | | | | | | | |
| Bloomfield | 50–100 | 93.31 | 89.13 | 96.96 | 91.95 | 8.18 | .497 |
| El Paso | 65–100 | 94.26 | 88.01 | 94.18 | 85.55 | 8.50 | .516 |
| Norwood | 60–100 | 95.29 | 91.50 | 98.79 | 90.20 | 7.85 | .477 |
| Oregon City | 50–100 | 89.31 | 80.58 | 95.77 | 84.55 | 6.70 | .407 |
| Portsmouth | 60–99 | 94.81 | 91.25 | 97.50 | 86.65 | 6.05 | .367 |
| Tupelo | 65–100 | 86.65 | 81.44 | 89.72 | 84.35 | 6.75 | .410 |

TABLE III (*Continued*)

III. READING PAPERS OR MAGAZINES

| | | | | | | | |
|---|---|---|---|---|---|---|---|
| Bloomfield ....... | 60–100 | 86.94 | 93.46 | 92.83 | 87.37 | 6.75 | .410 |
| El Paso .......... | 60–100 | 83.50 | 89.60 | 90.30 | 85.10 | 6.21 | .376 |
| Norwood ......... | 78–100 | 94.58 | 91.84 | 96.96 | 92.69 | 5.28 | .321 |
| Oregon City ...... | 65–100 | 90.56 | 83.93 | 96.05 | 89.89 | 9.09 | .552 |
| Portsmouth ...... | 72–100 | 95.11 | 91.92 | 97.32 | 92.87 | 6.09 | .552 |
| Tupelo .......... | 69–100 | 87.91 | 86.46 | 90.91 | 87.05 | 3.78 | .229 |

IV. INVESTING SAVINGS

| | | | | | | | |
|---|---|---|---|---|---|---|---|
| Bloomfield ....... | 60–100 | 96.10 | 90.82 | 98.15 | 91.76 | 8.55 | .519 |
| El Paso .......... | 75–100 | 93.52 | 89.25 | 95.53 | 90.80 | 6.03 | .366 |
| Norwood ......... | 78–100 | 96.92 | 94.73 | 98.29 | 95.12 | 4.44 | .269 |
| Oregon City ...... | 60–100 | 97.74 | 93.47 | 99.99 | 93.35 | 9.12 | .554 |
| Portsmouth ...... | 78–100 | 96.13 | 93.53 | 97.74 | 94.31 | 4.71 | .286 |
| Tupelo .......... | 78–100 | 88.74 | 86.34 | 92.88 | 87.69 | 3.60 | .218 |

V. WRITING LETTERS

| | | | | | | | |
|---|---|---|---|---|---|---|---|
| Bloomfield ....... | 78–100 | 95.03 | 93.11 | 94.14 | 93.68 | 3.30 | .200 |
| El Paso .......... | 72–100 | 90.55 | 87.51 | 92.80 | 88.97 | 6.88 | .356 |
| Norwood ......... | 90–100 | 97.73 | 96.61 | 98.86 | 96.74 | 1.95 | .128 |
| Oregon City ...... | 81–100 | 97.34 | 94.54 | 99.44 | 95.42 | 4.44 | .469 |
| Portsmouth ...... | 87–100 | 97.23 | 97.97 | 98.48 | 97.04 | 2.37 | .153 |
| Tupelo .......... | 81–100 | 97.28 | 96.70 | 98.49 | 96.68 | 2.43 | .147 |

VI. AND VII. FILLING OUT INCOME TAX AND TRAVELING

These last two life situations are so nearly uniform and the results from each city are so near to 100% non-use of all these problems that it is considered not necessary to list the results here in detail.

The correlations in Table IV have been computed on the basis of non-use, in per cents, of the 123 problems in each of the seven life situations listed on the Appendix B blank. For economy of space and convenience these life situations are numbered according to the Roman numerals of each situation listed under Table III, *i.e.*, I. Buying at the Store; II. Making Change; III. Reading; IV. Investing Savings; V. Writing Letters; VI and VII. Filling Out Income Tax and Traveling.

Data from the five cities listed below were correlated with the data from Portsmouth, Ohio, since Portsmouth had a larger number of cases (217) than any other one city.

TABLE IV

| | *I* | *P.E.* | *II* | *P.E.* | *III* | *P.E.* | *IV* | *P.E.* | *V* | *P.E.* | *VI* | *P.E.* |
|---|---|---|---|---|---|---|---|---|---|---|---|---|
| Bloomfield ... | .90 | .011 | 70 | .032 | 70 | .022 | 82 | .021 | 96 | .004 | 80 | .022 |
| El Paso ...... | .74 | .034 | 44 | .046 | 88 | .029 | 95 | .006 | 90 | .011 | 96 | .004 |
| Norwood ..... | .83 | .043 | 88 | .014 | 60 | .043 | 70 | .031 | 96 | .004 | 96 | .004 |
| Oregon City .. | .68 | .038 | 45 | .037 | 65 | .043 | 80 | .022 | 90 | .011 | 90 | .011 |
| Tupelo ....... | .75 | .036 | 56 | . 03 | 63 | .032 | 90 | .011 | 96 | .004 | 89 | .022 |

Correlation in column I, II, III in the above table have been re-checked for errors in calculation. Columns IV, V, VI have not been re-checked. Some of these correlations are the Pearson *r*. Others are the non-linear eta correlations.

### *I. Comments on Central Tendencies and Reliability of the Data*

One will readily see that there is great uniformity in the per cents of use and non-use in the data. Comparing the ages of those who filled out blank B we find that the range, median, Q1, and Q3 are fairly close in each of the cities studied. The schooling is sufficiently uniform to establish the fact that all those co-operating in filling out the blanks had covered the elementary school course of study in arithmetic.

More than three-fourths of those filling out blanks support or aid in the support of a family. Practically all check their bills from the store and 85 per cent of them check their bank statements. About 60 per cent of them are men, heads of families. This gives us evidence that we are dealing with a group roughly homogeneous in their use or non-use of arithmetic.

An examination of Table III indicates beyond question the approximate uniformity of the non-use of the 123 types of problems listed on blank B. Correlations of non-use of the 123 problems are sufficiently high to establish close relationship.

Garrett (*Statistics in Psychology and Education,* p. 298) states that, due to so many actual and potential sources of error in dealing in social data, the following coefficients of correlation denote substantial or high correlation:

*r* ranging from plus or minus .40 to .70 denotes substantial or marked correlation.

*r* ranging from plus or minus .70 to 1.00 denotes high correlation.

All the coefficients of correlation of the data here presented fall within the limits of plus or minus .40 to 1.00. We have, therefore, an incontestable case of high correlation.

### *II. Reliability*

Inspection of the data in Tables III and IV will show that the data given there meet all tests for reliability. (See Garrett, pp. 145-46.)

To offer another test of reliability and a test of "random sampling," the following technique was employed at the suggestion of Professor J. Ralph McGaughy. He used this technique to advantage in his study, *The Fiscal Administration of City School Systems* (pp. 69-74). He states that "if the critical ratio is as great as 4.5, the chances are almost perfect—9988 in 10,000—the true difference between the two obtained averages of a distribution is as indicated by the difference obtained."

To apply this technique to the data in six cities: The number of cases in blank B is 123; 25 per cent of these problems with the greatest non-use and 25 per cent of the problems which are the most used were selected from Bloomfield data, giving 31 cases. The average of the 31 problems was found both for the high end and the low end of the distribution. This, in one sense, made the selection of problems in Bloomfield a "select" group of problems. These same problems were listed according to the per cents of the greatest non-use and the most used in the other cities under consideration and the P.E. (difference of the averages) of Bloomfield and El Paso was computed. Each of the other cities in turn was thus compared with Bloomfield and with each other. The formula used:

$$P.E._{diff.}\ (A\ and\ B) = \sqrt{P.E.^2 av.\ A + P.E.^2 av\ B}$$

Data from this operation are given in Table V.

The interpretation of these critical ratios is that we can expect to find the actual difference in percentage of use of the problems by these cities just what the ratio indicates. In most cases this is not more than 5 per cent. In dealing with a problem of this kind, this result is as close as one could expect to get in determining the actual use of arithmetic by the common people.

Table V gives the P.E. (diff.) in these cities in the life situations "Buying at the Store" and "Reading," both for use and for non-use.

The elimination of all problems from the total of 345, the original list of problems, which are used by less than 25 per cent of those who checked the answeraires, leaves 44 problems. These are divided among the seven life situations as follows: Buying at the Store, 39 problems; Making Change, 5; Reading, 5; Writing, 1-5. The other life situations are not represented in problems used by 25 per cent of the population of these cities. The 5 used in Making Change are included in Buying. The problems used in all these situations combined are Problems No. 1, 2, 5, 7, 13, 14, 22, 25, 26, 27, 30, 31, 32, 35, 37, 38, 56, 58, 61, 67, 79, 82, 104,

TABLE V

BUYING AT THE STORE
NON-USE

| | P.E. (diff.) | D | Critical Ratio | Chances in 100 |
|---|---|---|---|---|
| Bloomfield and Portsmouth | 1.26 | 5.67 | 4.5 | 100 |
| Bloomfield and El Paso | 1.01 | 6.32 | 6.25 | 100 |
| Bloomfield and Tupelo | .94 | 13.04 | 14.40 | 100 |
| Bloomfield and Oregon City | .50 | 1.26 | 2.52 | 95 |
| Bloomfield and Norwood | .54 | 3.67 | 6.78 | 100 |
| Portsmouth and El Paso | 1.60 | .65 | .40 | 61 |
| Portsmouth and Tupelo | 1.35 | 7.37 | 5.46 | 100 |
| Portsmouth and Oregon | 1.34 | 6.93 | 5.17 | 100 |
| Portsmouth and Norwood | 1.36 | 2.00 | 1.47 | 84 |
| El Paso and Tupelo | 1.37 | 6.72 | 4.90 | 100 |
| El Paso and Oregon City | 1.10 | 7.58 | 6.88 | 100 |
| El Paso and Norwood | 1.13 | 2.75 | 2.41 | 95 |
| Tupelo and Oregon City | 1.05 | 14.30 | 13.62 | 100 |
| Tupelo and Norwood | 1.07 | 9.37 | 8.75 | 100 |
| Oregon City and Norwood | .71 | 4.93 | 6.94 | 100 |

BUYING AT THE STORE
MOST USED

| | P.E. (diff.) | D | Critical Ratio | Chances in 100 |
|---|---|---|---|---|
| Bloomfield and Portsmouth | 1.85 | 5.32 | 2.88 | 97 |
| Bloomfield and El Paso | 1.95 | 1.25 | .63 | 66 |
| Bloomfield and Tupelo | 1.95 | 1.81 | .93 | 74 |
| Bloomfield and Oregon City | 2.23 | 3.12 | 1.40 | 83 |
| Bloomfield and Norwood | 1.91 | 4.23 | 2.21 | 93 |
| Portsmouth and El Paso | 1.77 | 4.07 | 2.29 | 94 |
| Portsmouth and Tupelo | 1.88 | 7.13 | 2.25 | 93 |
| Portsmouth and Oregon City | 2.58 | 8.44 | 3.27 | 99 |
| Portsmouth and Norwood | 1.83 | .99 | .54 | 63 |
| El Paso and Tupelo | 2.58 | 3.06 | 1.18 | 78 |
| El Paso and Oregon City | 2.17 | 4.37 | 2.01 | 91 |
| El Paso and Norwood | 1.83 | 3.08 | 1.68 | 87 |
| Tupelo and Oregon City | 2.25 | 1.31 | .58 | 64 |
| Tupelo and Norwood | 1.93 | 6.14 | 3.18 | 98 |
| Oregon City and Norwood | 2.21 | 7.45 | 3.38 | 99 |

TABLE V—(*Continued*)

| | *P.E.* (*diff.*) | *D* | *Critical Ratio* | *Chances in 100* |
|---|---|---|---|---|
| | READING NON-USE | | | |
| Bloomfield and Portsmouth | .38 | .46 | 1.21 | 79 |
| Bloomfield and El Paso | .53 | 6.60 | 12.45 | 100 |
| Bloomfield and Tupelo | .38 | 5.83 | 15.34 | 100 |
| Bloomfield and Oregon City | .94 | 4.44 | 4.72 | 100 |
| Bloomfield and Norwood | .42 | .52 | 1.24 | 80 |
| Portsmouth and El Paso | .60 | 7.06 | 11.77 | 100 |
| Portsmouth and Tupelo | .45 | 6.29 | 13.98 | 100 |
| Portsmouth and Oregon City | .98 | 4.90 | 5.00 | 100 |
| Portsmouth and Norwood | .38 | .98 | 2.58 | 96 |
| El Paso and Tupelo | .59 | .77 | 1.31 | 81 |
| El Paso and Oregon City | 1.05 | 2.16 | 2.06 | 92 |
| El Paso and Norwood | .62 | 6.08 | 9.81 | 100 |
| Tupelo and Oregon City | .98 | 1.39 | 1.42 | 83 |
| Tupelo and Norwood | .50 | 5.31 | 10.62 | 100 |
| Oregon City and Norwood | 1.00 | 3.92 | 3.92 | 100 |
| | READING MOST USED | | | |
| Bloomfield and Portsmouth | 1.87 | 10.54 | 5.63 | 100 |
| Bloomfield and El Paso | 1.90 | 1.80 | .94 | 74 |
| Bloomfield and Tupelo | 1.63 | 7.94 | 4.87 | 100 |
| Bloomfield and Oregon City | 2.06 | 3.22 | 1.56 | 85 |
| Bloomfield and Norwood | 1.80 | 12.03 | 6.68 | 100 |
| Portsmouth and El Paso | 1.71 | 8.84 | 5.17 | 100 |
| Portsmouth and Tupelo | 1.40 | 2.70 | 1.92 | 90 |
| Portsmouth and Oregon | 1.91 | 7.42 | 3.88 | 100 |
| Portsmouth and Norwood City | 1.60 | 1.39 | .86 | 72 |
| El Paso and Tupelo | 1.34 | 6.14 | 4.58 | 100 |
| El Paso and Oregon City | 1.86 | 1.42 | .76 | 69 |
| El Paso and Norwood | 1.54 | 10.23 | 6.64 | 100 |
| Tupelo and Oregon City | 1.58 | 4.72 | 2.98 | 98 |
| Tupelo and Norwood | 1.18 | 4.9 | 4.15 | 100 |
| Oregon City and Norwood | 1.76 | 8.81 | 5.00 | 100 |

121, 139, 142, 145, 150, 153, 157, 162, 163, 170, 176, 181, 202, 214, 215, 216, 217, 238, 239, 240, 319.

Without attempting to work out any elaborate formula to show relative values of the 44 problems listed above, the writer has used "Per cent position equals $\frac{100\ (R—.5)}{N}$" (Garrett, *Statistics in Psychology and Education,* p. 113) and a table derived from this percentage formula to get the data for Table VI below. The formula is used in the following way: All the 39 problems listed under Buying were arranged in a descending order of average percentage of use in the six cities in Buying. This table we are calling the *Social Value Index*. It is intended to show the relative importance, or value, of the problems used. It is assumed that this index shows the value of the problems because it is based

TABLE VI

| I<br>Problem No. | II<br>Av. Use in Six Cities | III<br>Rank of Problem Based on Use | IV<br>Percent Position | V<br>Social Value Index | VI<br>Category Under Which Problem Occurs |
|---|---|---|---|---|---|
| 13 | 58 | 1 | 1.3 | 9.0 | Dry Measure |
| 14 | 53 | 2 | 3.8 | 8.4 | Dry Measure |
| 26 | 50 | 3 | 6.4 | 8.0 | Dry Measure |
| 67 | 48 | 4 | 8.9 | 7.7 | Long Measure |
| 162 | 46 | 5 | 11.5 | 7.4 | Fractions in C. D. |
| 145 | 45 | 6 | 14.1 | 7.0 | Simple Fractions |
| 157 | 44 | 7 | 16.6 | 6.9 | Fractions |
| 35 | 43 | 9 | 21.8 | 6.5 | Dry Measure |
| 142 | 43 | 9 | 21.8 | 6.5 | Simple Fractions |
| 238 | 43 | 9 | 21.8 | 6.5 | Cash Discount |
| 31 | 42 | 11 | 26.9 | 6.2 | Avoirdupois |
| 153 | 41 | 12 | 29.4 | 6.1 | Fractions |
| 27 | 40 | 14 | 34.6 | 5.8 | Avoirdupois |
| 30 | 40 | 14 | 34.6 | 5.8 | Compound Denominate Numbers |
| 139 | 40 | 14 | 34.6 | 5.8 | Simple Fractions |
| 150 | 39 | 16.5 | 41.0 | 5.5 | Simple Fractions |
| 170 | 39 | 16.5 | 41.0 | 5.5 | U. S. Money |
| 25 | 37 | 18 | 44.8 | 5.3 | Dry Measure |
| 38 | 36 | 19.5 | 48.7 | 5.3 | Cubic Measure |
| 58 | 36 | 19.5 | 48.7 | 5.3 | Dry Measure |
| 82 | 35 | 21.5 | 53.8 | 4.7 | Liquid Measure |
| 239 | 35 | 21.5 | 53.8 | 4.7 | Store Discount |
| 56 | 34 | 23.5 | 58.9 | 4.7 | Paper by Quire |
| 79 | 34 | 23.5 | 58.9 | 4.7 | Dry Measure |
| 319 | 33 | 25 | 62.8 | 4.3 | Square Measure |
| 61 | 32 | 27 | 67.9 | 4.1 | Liquid Measure |
| 176 | 32 | 27 | 67.9 | 4.1 | U. S. Money |
| 215 | 32 | 27 | 67.9 | 4.1 | Aliquot Parts |
| 32 | 30 | 29.5 | 74.3 | 3.7 | Avoirdupois |
| 181 | 30 | 29.5 | 74.3 | 3.7 | Dry Measure |
| 163 | 28 | 31 | 78.2 | 3.5 | Fractions Div. |
| 37 | 27 | 34 | 85.8 | 2.9 | Cubic Measure |
| 121 | 27 | 34 | 85.8 | 2.9 | Small Fractions |
| 202 | 27 | 34 | 85.8 | 2.9 | Decimals |
| 216 | 27 | 34 | 85.8 | 2.9 | Aliquot Parts |
| 240 | 27 | 34 | 85.8 | 2.9 | Store Discount |
| 104 | 26 | 37 | 93.5 | 2.0 | Liquid Measure |
| 217 | 25 | 38.5 | 97.4 | 1.2 | Decimals |
| 214 | 25 | 38.5 | 97.4 | 1.2 | Aliquot Parts |

upon the relative universality of use among those filling out the blanks. A problem that is used by 58 people out of a hundred, as is the case of No. 13, for example, is "relatively" more important —valuable—than one that is used by only five people out of a hundred.

Column VI, the category under which the problem is classed, requires some explanation. Repeated "Dry Measure" and other forms of "Compound Denominate Numbers" mean that these are used in different connections; for example, subtraction of fractional parts of measures. Most of the discount categories involve percentage without time, such as buying goods at cash discount of certain amounts. From a study of the Social Value Index in the above table it is clear that problems dealing with Compound Denominate Numbers are much more valuable to those checking the blanks than any other category listed. This implies, of course, that fractional parts of units of measurement are involved. Practically all the fractions that are used, both decimal and simple, are included in this category. United States Money requires nearly all the decimals that are used.

Problems in Dry Measure and Long Measure are about nine times more valuable than problems in Decimal Fractions, especially when decimals are used in any other relation than that needed in United States Money.

The original list of problems included eighteen categories. These are listed on page 26. On the basis of eliminating all problems used by less than 25 per cent of those checking, only eight categories remain. These are:

1. Numeration and Notation used chiefly in reading.
2. Compound Denominate Numbers.
   *a*. United States Money.
   *b*. Dry Measure.
   *c*. Liquid Measure.
   *d*. Avoirdupois.
   *e*. Square Measure.
   *f*. Cubic Measure.
   *g*. Paper Measure—the quire or ream.
3. Simple Fractions.
   *a*. Addition.
   *b*. Subtraction.
   *c*. Multiplication and Division.

4. Decimal Fractions used in United States Money, etc.
5. Percentage in its simplest form and only in one case.
6. Simple Store Discount.
7. Aliquot Parts of 100.
8. Mensuration.

Of these eight categories the following are the most important as shown by the Social Value Index given above:

1. Compound Denominate Numbers.
2. Simple Fractions.
3. Cash Discount at the store or percentage without the element of time.

If we use the 345 problems which appear on blank A as a basis for calculating the percentage of those problems that are used (44 in all), we find that those who filled out the blanks used only 12½ per cent of all problems listed.

# CHAPTER VI

## GENERAL SUMMARY AND CONCLUSIONS

Of the 345 problems on the first "answeraire," A, certain problems were eliminated. Those problems on which there was some variation of use were included in the second blank. Data collected from this second blank, B, agree with those of the first blank on the problems to be eliminated, with the exception of sixteen; that is, the types of problems on both blanks which are eliminated are the same, with the exception of these: Nos. 18, 19, 23, 24, 40, 226, 235, 244, 246, 248, 262, 265, 280, 325, 343, 344.

In the first "answeraire" there are 345 type problems classified under headings called categories. There are eighteen of these categories, such as Compound Denominate Numbers, Fractions, etc. It is arbitrarily assumed that problems which are or have been used by less than 25 per cent of those filling out the blanks should be discarded. On this basis of elimination, only eight out of the eighteen categories are left.

It should be noted that, with few exceptions, the only life situations under consideration in which any of the above categories are used are Buying and Reading. These few exceptions are in Making Change and in Writing.

Of the 345 problems listed in the first blank, A, only 44 are used.

More than 85 per cent of those filling out the blank stated that they checked their bills from the store and their monthly bank statements. Arithmetic processes necessary for checking bills and bank statements are chiefly subtraction, multiplication, and simple additions.

The Social Value Index gives a basis for evaluating the importance of the 44 problems remaining after the elimination. It shows that Compound Denominate Numbers with their variations is the most important category in the eight categories remaining after the elimination. Simple Fractions used in connection with

Compound Denominate Numbers and with United States Money and in a few other situations is the next most important category. Simple Operations in Percentage without the time element is the third most important.

The historical material clearly indicates that we have not arrived at the determination of the content of our public school arithmetic by any very scientific technique. Even textbook writers have not solved the problem to their own satisfaction. There is wide variation in the quantity and kind of problems which they offer in their texts. If we are to be governed by the type and amount of arithmetic included in *all* the texts to-day, we certainly see that we are not *much* nearer a reduction of the amount of required arithmetical content than we were in 1900.

*I*

This study not only gives us a basis for selecting the minimum content but also furnishes a method of evaluating its importance. It shows where the emphasis should be placed.

*II*

The chapter on the "Presentation of the Data" proves that not more than 60 per cent of those who checked the blanks use little, if any, arithmetic beyond the four fundamental operations.

*III*

If we assume that the problems used in this study are typical, we are safe in saying that the schools have been teaching more than 85 per cent more arithmetic than is required for the accomplishment of the values which we have described under the life situations: Buying at the Store, Making Change, Reading, Investing One's Savings, Writing Letters, etc., and Traveling, etc.

# CHAPTER VII

## SOME APPLICATIONS OF THE RESULTS OF THE PRESENT STUDY TO SCHOOL USES

### *I*

The universal use, by all literate adults, of the four fundamentals, addition, subtraction, multiplication, and division of simple whole numbers is assumed.

### *II*

It has been arbitrarily taken for granted that less than 25 per cent use by the adult population of the United States warrants our discarding certain types of problems from elementary school textbooks.

### *III*

This basis of elimination facilitates the selection of the types of material that should form the basis of any arithmetic textbook written for use in the elementary school.

Since this study has shown that in the life situations which we have considered, the problems which are used—chiefly in buying, in making change, and in general reading—are of a simple kind, it furnishes a basis for revising the arithmetic course of study. Not only does it give us a basis for selecting the *content,* but it also provides a clue to a basis for working out a method of presenting the material. It indicates where the emphasis should be placed on certain problems. For example, those problems that are chiefly used in reading need not be given the thorough drill that is given those which are chiefly used in buying. The reading basis could easily be used to give informational arithmetic.

### *IV*

If the schools are teaching 85 per cent more arithmetic than is necessary to realize the values which we are considering in this study, we have a means of solving one of the problems of non-promotion in the elementary schools. This enters not only as a factor in school instructional efficiency, but also in school costs. Failures in school are expensive to the tax-paying public.

It is not only expensive to have failures but depressing and injurious to a growing personality.

### *V*

Elimination of unsuitable and ill-placed school work aids the interest factor in a student's life.

The elimination of useless arithmetic material gives more space for the inclusion of other materials that may be found to be socially valuable but for which we have no time in the school day at the present time.

### *VI*

Arithmetic has been shown to be the chief source of non-promotions in the elementary schools. (Judd, *Measuring the Work of the Public Schools,* in *Cleveland Survey,* pp. 94-123.) It is too abstruse and couched in language beyond the years of experience and comprehension of most pupils who are required to study it.

### *VII*

To know the amount of arithmetic used aside from vocational needs is as important as to know what is needed for vocational use,—if not more important. Not all people follow or need the vocational arithmetic used in any one vocation, but everyone needs approximately the arithmetic represented by the findings of this study.

### *VIII*

Perhaps the method used in this study, although it is not altogether new, may give suggestions which will help in formulating a technique for the solution of other curricular problems in other elementary school subjects.

# CHAPTER VIII

## SOME UNSOLVED PROBLEMS ARISING FROM THE PRESENT STUDY

It is probably true that the solution of any problem of importance discloses as many problems as it solves; particularly is this likely to be the case in important social problems which deal with social values and relationships. Many problems of this nature are so subtle that they escape most of us until the disharmony resulting from them forces itself upon our attention. There are several interesting corollaries in the nature of problems that present themselves.

### *I*

The proper next step lying within the field of this study is that of further refinement of educational objectives in arithmetic. The present investigation has attempted mainly to determine what is used and what is not useful in arithmetic in social life. It has not been the aim of this investigation to go into further detail as to *how much* of the types of arithmetic found here to be useful. It is evident that some of the arithmetic that is useful is not as useful as other types. Some future study may well be based upon the determination of the *degrees* of usefulness of certain kinds and amounts of arithmetic that have been found in this investigation to be valuable.

### *II*

This investigation has attempted to bring vocational arithmetic more fully into relief. As was stated in Chapter II, when the division between vocational and social needs in arithmetic has been more clearly drawn, analyses of the amount and kind of arithmetical needs in each vocation may be more clearly stated and the solution of the vocational problem proved to be relatively simple.

## *III*

Another problem which should be easily isolated relates to that type known as "informational arithmetic." The question often arises, "If we eliminate much of the arithmetic now in the course of study would it not be necessary to teach a great deal of informational arithmetic?" This problem is partly covered in one of the life situations in the present study, under the heading Reading; that is, what kind of arithmetical knowledge does one need in reading newspapers, magazines, etc., intelligently? A checking over of the newspapers and magazines of the 682 taking part in this study probably reveals but little use of any kind of arithmetic now taught in the public schools. There is needed, however, a kind of arithmetical knowledge, relating to economic activities of county, state, or nation, banking, stocks, and bonds, etc., which may well be grouped into proper instructional units. No one at present knows just what the kind or amount of such knowledge is or what it should be. A method may be worked out on the basis of an analysis of the actual knowledge of this informational type necessary for one to understand the several grades of newspaper and magazine reading. It would require an immense amount of checking of such periodicals to make the basis of comparisons.

There is another kind of informational mathematics that may be found to be generally useful, such as some bits of the history of mathematics for purely cultural purposes, some bits of geometry and other elementary forms of mathematics which for informational purposes could well be included in the public school course.

Buswell and Judd (*Investigations Relating to Arithmetic,* p. 158) have this to say concerning the inclusion of certain types of mathematics in the elementary school program: "There is certainly no good historical reason why the teaching of mathematics in the elementary grades should not include exercises in the concrete aspect of geometry." They do not propose that the school attempt to instruct little children in the abstract phases of demonstrative geometry. We are assured that much could be included in this form of mathematics. No one knows just how much of this sort of thing is valuable at the present time. They state in another place (*Ibid.,* p. 17): "The courses in arithmetic as now organized include, in addition to number experiences, a great many ideas which are intelligible only when pupils are taught some of the facts of social life. It is especially true of denominate numbers that they can be understood and properly used only when their use is accompanied by much instruction which is not at all arithmetical in character." But there is still another phase of informational arithmetic that could well be taught in the public schools which would be generally useful—probably much more so than much that is now taught. This phase has not been well defined, neither has any one taken the pains to ferret out the extent of mathematical or semi-mathematical knowledge that could be rather generally cultural for given classes in adult populations. There are several published attempts at finding out these different kinds of informational arithmetic. For example, the one by Alice Camerer, referred to in a previous chapter. There are other investigations that touch this subject but none are thorough enough to warrant any far-reaching conclusions.

Along with an investigation of what and how much informational arithmetic to include in the arithmetic course should also go a parallel investigation to determine the grade-placement of the material and methods of teaching it. We could hope by such a study to include large amounts of informational problems and situations and in this way secure actual and more intelligent application of arithmetic in life situations, that is, more arithmetic in the productive work of the kitchen, in intelligent buying, in correct farm and home accounting, and in intelligent saving and investing. Textbook makers have been active in such investigations and have included some of the findings in textbooks for the schools.

## *IV*

Once the amount and kind of arithmetic content has been determined, there arises the question of how long elementary pupils should study this content. The Committee of Fifteen gave the following statement regarding the amount of time necessary to devote to arithmetic in the elementary schools: "Your committee believes that, with the right method and a wise use of time in preparing the arithmetic lesson in and out of school, five years are sufficient for the study of mere arithmetic—the five years beginning with the second school year and ending with the close of the sixth year." It would seem that from the information contained in this investigation even five years is too much time to give to the subject, and that being the case, when should we begin formal arithmetic? If considerable time is saved by shortening the course in arithmetic, the ramifying spread of such a change in the present requirements would affect all other subjects now taught. Many questions will thus arise and demand solution, once we have accepted the conclusions of the present investigation.

Some studies have been made on this point, and it has been shown that pupils who have not started arithmetic before the second grade have been able to advance equally as well and as fast as those who began in the first grade, and in many cases with greater satisfaction. "One writer gives an account of an experiment in New York City of the complete omission of arithmetic in Grades IB and IA. The time thus saved was devoted to reading. The results showed that pupils reached the usual level of attainment in arithmetic by the end of Grade IIB and apparently had suffered in no way from the omission of arithmetic." (Buswell and Judd, p. 19.) The question is now proper: May not pupils, generally, in the elementary schools acquire such arithmetical tools and concepts as are needed, if instruction in arithmetic be deferred to the fourth or fifth grade? No one knows the answer.

## *V*

Then, there arises the important question with all of its ramifying implications—if the amount of arithmetic is materially lessened, what may properly occupy the time of the pupil and the space in the curriculum formerly used in teaching the great amount of arithmetic?

## *VI*

Perhaps elementary school curricula should spread out laterally, that is, include a wider range of subjects, or include more of the total range of human culture than they do at present. But the greater the expansion in this way the less thoroughly any one subject may be taught. This brings up the question of *how* thoroughly is *thoroughly enough?* It will be remembered that the elements, the beginnings, of all fields of human knowledge are simple—so simple that any normal mind, and that, too, of boys and girls as well as adults, may readily grasp it. More of the "spiritual" side of human culture could be included—using the word "spiritual" in the sense which Professor David Snedden uses the term. But what and how much of the "spiritual?"

## *VII*

The reading of *graphs* and similar devices dealing with concepts of quantity and trends which are found in the literature of to-day is growing in importance as a social "need." The solution of this problem lies wholly in the future. No one knows what kinds of graphs are used, or to what extent. A survey of current reading matter would be one of the steps in a solution.

## *VIII*

Are there really any child needs in arithmetic that adults do not use in some way or other? This has not been determined, although there are those who claim there is a difference. Buswell and Judd have added this problem to the above list (p. 21). Numerous further studies of the curriculum will have to be made before a well-organized course in arithmetic can be planned with scientific assurance. What has been done up to this time opens the way to analytical study but can hardly be said to have determined, with anything like completeness, the proper content and arrangement of the topics of the course.

Many more problems concerning the quantity, kind, and placement of the materials of instruction in arithmetic could be added to this list, but these are enough to indicate that the field is wide and but little real scientific work has yet been done.

# APPENDIX A

## "ANSWERAIRE" A

NAME____________________________________AGE?__________
Do You Support a Family?_________Grade at Which You Left School?____
Occupation?__________________ Usual Magazines and Newspapers Read:
(1)____________________________________________
(2)____________________________________________
(3)____________________________________________
(4)____________________________________________

### DIRECTIONS FOR FILLING OUT THE "ANSWERAIRE"

In the following pages indicate by placing a check (√) in the proper column which describes the frequency of your use of the problems on the left of the sheet in Making Purchases, Making Change, Reading, Investing, Writing, and Travel.

1. Underscore the word indicating the frequency with which you check with your monthly bank statements to see if they are correct:
   Very often.
   Often.
   Occasionally.
   Seldom.
   Very seldom.
   Never.

2. Underscore the word indicating the frequency with which you check with your monthly grocery, dry goods, etc., bills to determine if they are correctly charged or added:
   Very often.
   Often.
   Occasionally.
   Seldom.
   Very seldom.
   Never.

[The problems in the "Answeraire" are given on pages 56 to 66.]

| In this column are given types of problems | BUYING | | | | | |
|---|---|---|---|---|---|---|
| | *Very often* | *Often* | *Occasionally* | *Seldom* | *Very seldom* | *Never* |
| 1. Express the following in figures: Two million, five hundred twenty-two thousand, nine hundred and sixty-three. | | | | | | |
| 2. Express the following in figures: Thirty-five billion. | | | | | | |
| 3. Put the following expression in equivalent figures: Thirty trillion. | | | | | | |
| 4. Read the following: $3,208,035.00. | | | | | | |
| 5. Read the following: $320,803,538.00. | | | | | | |
| 6. Read the following: $1,257,865,783,375.00. | | | | | | |
| 7. XV. | | | | | | |
| 8. LXXVIII. | | | | | | |
| 9. MDCCCCLXXXVIII. | | | | | | |
| 10. Reduce $3.00 to mills. | | | | | | |
| 11. Reduce 10,375 mills to dollars. | | | | | | |
| 12. Reduce 1 Eagle to ¼ dollars. | | | | | | |
| 13. At $1.75 per yard what will 9 inches of silk cost? | | | | | | |
| 14. At 25c per foot what will 3 rods of wire fence cost? | | | | | | |
| 15. If 1 mile of a railroad is worth $400,000, what is one yard worth? | | | | | | |
| 16. What is the value of the Peseta in United States money? | | | | | | |
| 17. What is the value of the Lire in United States money? | | | | | | |
| 18. What is the value of the Crown in United States money and in all countries where the Crown is used? | | | | | | |
| 19. What is the value of the Yen in United States money? | | | | | | |
| 20. What is the value of the Tael in United States money? | | | | | | |
| 21. What is the value of the Peso in United States money? | | | | | | |
| 22. What is the value in United States money of the English pound sterling, £? The shilling? | | | | | | |

| MAKING CHANGE | | | | | | READING | | | | | | INVESTING | | | | | | WRITING | | | | | | TRAVEL | | | | | |
|---|---|---|---|---|---|---|---|---|---|---|---|---|---|---|---|---|---|---|---|---|---|---|---|---|---|---|---|---|---|
| Very often | Often | Occasionally | Seldom | Very seldom | Never | Very often | Often | Occasionally | Seldom | Very seldom | Never | Very often | Often | Occasionally | Seldom | Very seldom | Never | Very often | Often | Occasionally | Seldom | Very seldom | Never | Very often | Often | Occasionally | Seldom | Very seldom | Never |
| | | | | | | | | | | | | | | | | | | | | | | | | | | | | | |
| | | | | | | | | | | | | | | | | | | | | | | | | | | | | | |
| | | | | | | | | | | | | | | | | | | | | | | | | | | | | | |
| | | | | | | | | | | | | | | | | | | | | | | | | | | | | | |
| | | | | | | | | | | | | | | | | | | | | | | | | | | | | | |
| | | | | | | | | | | | | | | | | | | | | | | | | | | | | | |
| | | | | | | | | | | | | | | | | | | | | | | | | | | | | | |
| | | | | | | | | | | | | | | | | | | | | | | | | | | | | | |
| | | | | | | | | | | | | | | | | | | | | | | | | | | | | | |
| | | | | | | | | | | | | | | | | | | | | | | | | | | | | | |
| | | | | | | | | | | | | | | | | | | | | | | | | | | | | | |
| | | | | | | | | | | | | | | | | | | | | | | | | | | | | | |
| | | | | | | | | | | | | | | | | | | | | | | | | | | | | | |
| | | | | | | | | | | | | | | | | | | | | | | | | | | | | | |
| | | | | | | | | | | | | | | | | | | | | | | | | | | | | | |
| | | | | | | | | | | | | | | | | | | | | | | | | | | | | | |
| | | | | | | | | | | | | | | | | | | | | | | | | | | | | | |
| | | | | | | | | | | | | | | | | | | | | | | | | | | | | | |
| | | | | | | | | | | | | | | | | | | | | | | | | | | | | | |
| | | | | | | | | | | | | | | | | | | | | | | | | | | | | | |
| | | | | | | | | | | | | | | | | | | | | | | | | | | | | | |
| | | | | | | | | | | | | | | | | | | | | | | | | | | | | | |

Problems 23 to 345 were arranged on the blanks in the same manner as problems 1 to 22 above.

23. What is the value in United States money of the franc, centime, etc.?
24. Do you use the German mark, pfennig, thaler?
25. At 5c per pint, what will 3 bu. of rice cost?
26. If one peck of peaches sells for $1.00, what will 1 pint cost?
27. If 8 qt. of flour cost $2.25, what will 1 bu. 2 qt. and 1 pt. cost?
28. 1 gill of cider cost 5c; what will 1 gallon cost?
29. 1 gal. of salt water weighs 11 lb.; what will be the weight of 3 qt. 2 pt. and 3 gills?
30. At $1.00 per gal., what will 2 gal. 3 qt. 1 pt. and 3 gills of vinegar cost?
31. 2 cwt. of sweet potatoes equals how many pounds?
32. What is the weight of 180 iron castings each weighing 75 lbs.?
33. What will be the cost of 1 T. 6 cwt. 4 lb. 2 oz. of coal @ 4c per pound?
34. What will 1 sq. mi. of land cost @ $1.00 per sq. ft.?
35. What will be the cost of 3888 sq. in. of carpet at $5.00 per sq. yd.?
36. A man bought 515,280 sq. rd. of land. How many sq. mi. did he buy?
37. Reduce 2 cu. yd. to cubic inches.
38. In a pile of wood 40 ft. long, 12 ft. wide, and 5 ft. high, how many cords?
39. What will be the cost of excavating a cellar 24 ft. long, 13 ft. wide, and 6 ft. deep, at $1.25 per cu. yd.?
40. Reduce 2 hr. to seconds.
41. Reduce 422,000 sec. to days.
42. Reduce 41,761 min. to months.
43. Reduce 5 lbs. 4 oz. Troy to ounces.
44. Reduce 12,530 grams to pounds Troy.
45. Reduce 8 lbs. 9 oz. 13 pwt. 17 grams to grams.
46. Reduce 4 lb. 5℥ 2 grams to grams.
47. Reduce 431 ℥ to pounds.
48. Reduce 6321 grams to pounds.
49. Reduce 3 yd. to Barleycorns.
50. How many lines in 1 ft. 6 in.?
51. Reduce 1 span 4 in. 3 barleycorns to lines.
52. Reduce 8° 41′ 45″ to seconds.
53. Reduce 3661″ to degrees.
54. Reduce 915′ to degrees.
55. How many sheets of paper will be required for a 12 mo. book of 336 pages?
56. At 18c a quire, what will 3 bundles of paper cost?
57. How many reams of writing paper in 1000 quires?
58. Add: 4 bu. 3 pk. 7 qt. and
    5 bu. 2 pk. 2 qt.
59. Five loads of wheat measured thus: 21 bu. 3 pk.; 14 bu. 1 pk.; 23 bu. 2 pk.; 18 bu. 1 pk.; 22 bu. 1 pk. How many bushels in all?
60. Add: 13 lb. 11 oz.; 17 lb. 13 oz.; 14 lb. 14 oz.; 16 lb.; 19 lb. 7 oz.; and 17 lb. 9 oz.
61. Add: 7 qt. 1 pt. 3 gills and 6 qt. 0 pt. 2 gills.

62. A grocer bought at different times cider as follows: 40 gal. 3 qt. 1 pt. 3 gills. Another time, 71 gal. 2 qt. 1 pt. 2 gills. How much cider in all?
63. Add: 3 gal 2 pt. and 3 qt. 3 gills.
64. Add: 45 T. 3 cwt. and 53 lb. 10 oz.
65. Add: 16 T. 14 cwt. 75 lb. 15 oz and 19 T. 17 cwt. 18 lb. 13 oz.
66. A grocer sold 5 hogsheads of sugar. The 1st weighed 8 cwt. 36 lb.; the 2nd, 4 cwt. 64 lb.; the 3rd, 5 cwt. 19 lb.; the 4th, 7 cwt. 84 lb.; the 5th, 7 cwt. 84 lb. What did all weigh?
67. Add: 2 ft. 3 in. and 13 ft. 9 in.
68. Add: 28 mi. 129 rd.
    64 mi. 280 rd.
    17 mi. 275 rd.
69. Two men depart from the same place; one travels 104 mi. 50 rd. due east; the other 95 mi. 270 rd. due west. How far are they apart?
70. Add: 15 sq. yd. 8 sq. ft. 115 sq. in.
    20 sq. yd. 8 sq. ft. 109 sq. in.
71. Add: 41 A. 51 sq. rd. 14 sq. yd. 5 sq. ft.
    193 A. 155 sq. rd. 20 sq. yd. 8 sq. ft.
72. A farmer has three farms; in the 1st are 186 A. 134 sq. rd.; in the 2nd, 286 A. 17 sq. rd.; in the 3rd, 113 A. 89 sq. rd. 3 sq. yd. How much has he in all?
73. Add: 50 cu. yd. 18 cu. ft. 900 cu. in.
    46 cu. yd. 20 cu. ft. 990 cu. in.
74. Add: 13 cord 28 cu. yd. 45 cu. ft. 18 cu. in.
    20 cord 67 cu. yd. 30 cu. ft. 9 cu. in.
75. A has 4 piles of wood; in the 1st, 7 cords 78 cu. ft.; in the 2nd, 16 cords 24 cu. ft.; in the 3rd, 35 cords 127 cu. ft.; in the 4th, 29 cords 10 cu. ft. How much in all?
76. Add: 16 days 18 hr.
    1 day 3 hr.
77. Add: 13 days 15 hr. 49 min. 59 sec.
    19 days 16 hr. 53 min. 42 sec.
78. Add: 3 mon. 2 wk. 4 days 19 hr. 30 min. 3 sec.
    1 mon. 1 wk. 5 days 13 hr. 27 min. 18 sec.
79. From—5 bu. 3 qt.
    take 2 bu. 2 qt.
80. From—3 bu. 3 pk. 3 qt.
    take 2 bu. 1 pk. 1 qt.
81. From—12 bu. 0 pk. 1 qt. 0 pt.
    take 8 bu. 2 pk. 1 qt. 1 pt.
82. From—17 gal. 2 qt.
    take 13 gal. 3 qt.
83. From—18 gal. 2 qt. 1 pt.
    take 10 gal. 3 qt. 0 pt.
84. From—43 gal. 1 qt. 1 pt. 2 gi.
    take 23 gal. 3 qt. 2 pt. 3 gi.
85. From—12 cwt. 50 lb.
    take 10 cwt. 75 lb.

86. From—14 T. 18 cwt. 75 lb.
    take 9 T. 20 cwt. 80 lb.
87. From—16 T. 7 cwt. 18 lb. 14 oz.
    take 5 T. 6 cwt. 75 lb. 15 oz.
88. From—18 mi. 198 rd.
    take 11 mi. 236 rd.
89. From—10 mi. 200 rd. 4 yd.
    take 8 mi. 250 rd. 5 yd.
90. From—20 mi. 230 rd. 5 yd. 1 ft. 3 in.
    take 15 mi. 40 rd. 10 yd. 2 ft. 7 in.
91. From—19 sq. yd. 6 sq. ft.
    take 17 sq. yd. 7 sq. ft.
92. From—327 A. 148 sq. rd. 19 sq. yd.
    take 77 A. 155 sq. rd. 25 sq. yd. 1 sq. ft.
93. From—200 A. 80 sq. rd. 1 sq. yd. 3 sq. ft. 100 sq. in
    take 100 A. 90 sq. rd. 9 sq. yd. 4 sq. ft. 140 sq. in.
94. From—18 cu. yd. 7 cu. ft. 0 cu. in.
    take 9 cu. yd. 8 cu. ft. 3 cu. in.
95. From—28 cord 116 cu. yd. 116 cu. ft.
    take 19 cord 119 cu. yd. 120 cu. ft.
96. From—30 cord 27 cu. yd. 120 cu. ft. 8 cu. in.
    take 15 cord 28 cu. yd. 125 cu. ft. 100 cu. in.
97. From—18 hr. 43 min. 27 sec.
    take 17 hr. 51 min. 45 sec.
98. From—3 wk. 20 da. 17 hr. 40 min.
    take 1 wk. 29 da. 23 hr. 58 min.
99. From—1 mo. 2 wk. 15 da. 20 hr. 50 min.
    take 3 wk. 28 da. 15 hr. 58 min.
100. How much hay in 7 loads each weighing 10 cwt. 87 lb.?
101. Multiply 23 cu. yd. 9 cu. ft. 228 cu. in. by 12.
102. If a ship sails 208 mi. 176 rd. a day, how far will it sail in 15 days?
103. Multiply 2 wk. 4 da. 13 hr. 48 min. 39 sec. by 75.
104. Multiply 47 gal. 3 qt. 1 pt. by 59.
105. Multiply 3 A. 252 sq. rd. 10 sq. yd. 28 sq. ft. 109 sq. in. by 39.
106. Divide 57 bu. 3 pk. 4 qt. 1 pt. among 5 persons.
107. I travel 39 mi. 288 rd. in 17 hr. At what rate per hour do I travel?
108. Divide 69 A. 64 sq. rd. by 16.
109. Divide 45 T. 18 cwt. 75 lb. by 17.
110. Divide 10 gal. 3 qt. 1 pt. 3 gi. by 3.
111. Divide 114 da. 22 hr. 45 min. 18 sec. by 54.
112. When it is noon in Cincinnati, what is the time in Philadelphia?
113. The difference of longitude between two places is 30 degrees; what is the difference in time?
114. The difference in time between New York and St. Louis is 1 hr. 4 min. 56 sec. What is the difference in longitude?
115. What are the prime factors of 2310?
116. What prime factors are common to 60 and 90?
117. What prime factors are common to 46, 69, 92 and 115?
118. What is the least common multiple of 4, 6, and 8?

119. Find the least common multiple of 120, 360, 144, 720 and 72.
120. Find the least common multiple of 42, 56, and 63.
121. In 3 apples, how many fourths?
122. In 8 ft., how many twelfths?
123. In 25 bushels, how many twentieths?
124. In 4 and one-tenth, how many tenths?
125. In 5 and three-fourths, how many fourths?
126. Reduce 21 117/583 to an improper fraction.
127. In 17/5 of a dollar, how many dollars?
128. Divide $3000 equally among 77 men.
129. Divide 1325 lb. of sugar equally among 101 men.
130. Divide ⅞ dollars equally among 72 boys.
131. Divide 3/5 of a ton of coal equally among 60 families.
132. Give an equal share of 9/10 of a dollar to each of 100 children.
133. Reduce 18/60 to its lowest terms.
134. Reduce 873/1067 to its lowest terms.
135. Reduce 6465/7335 to its lowest terms.
136. Reduce ½, 2/3, ¾ to their least common denominator.
137. Reduce 7/16, 11/18, 17/24, 19/36, 25/48, to their least common denominator.
138. Reduce 3/5, 7/10, 6/25, 11/30, 13/45, 23/60 to their least common denominator.
139. Add: ¼, 2/4, ¾.
140. Add: 5/13, 8/13, 9/13, 11/13.
141. Add: 12/25, 16/25, 18/25, 24/25.
142. Add: 1/2 and 5/6.
143. Add: 1/12, 1/13, 1/14, 1/15.
144. Add: 1 1/3, 1¼, 3½.
145. Subtract ¼ from ¾.
146. From 8/10 subtract 3/10.
147. From 3⅛ subtract 1⅜.
148. Subtract 1/3 from 1/2.
149. From 4/15 subtract 1/10.
150. From 3 1/2 subtract 1 2/3.
151. If each of 8 men earns 2/3 dollars per hr. for work, how much do they earn in 1 hr. working all together?
152. Multiply 8/13 by 11/16.
153. Multiply 2¼ by 3½.
154. Multiply together 5/12, 9/16, 2 2/11.
155. Multiply together 2 1/5, 2 3/6, 3 4/5.
156. Multiply together 2 2/3, 6 2/5, 3 1/4, 7/13.
157. What is ¾ of 5?
158. What is 2/3 of 3/4 of 4/5?
159. What is 7/12 of 28?
160. What will be the cost of 2 1/3 lb. of meat at 33 1/5 cents a lb.?
161. At the rate of 5½ miles an hour, how far will a man travel in 7¾ hrs.?
162. What is the cost of 1 2/3 yd. of muslin at $3/20 per yard?
163. Divide 2 2/5 by 6.
164. Divide 130 2/3 by 18.

165. Divide 14/15 by 21.
166. Divide 3/5 of 8/9 by 6/7 of 3/4.
167. Divide 5/18 of 2/5 of 12 1/5 by 1/5 of 8 1/5.
168. Divide 2/7 of 7/8 by 3/4 of 1/3 of 5.
169. 3 days is what part of a week?
170. ¼ of a dollar is what part of ½ dollar?
171. 21/32 is what part of 34/48?
172. Reduce $\frac{3/4}{2/3}$ to a simple fraction.
173. Reduce $\frac{7\ 2/3}{8\ 3/4}$ to a simple fraction.
174. Multiply $\frac{1\frac{1}{2}}{2\ 1/3}$ of $\frac{3/7}{4\frac{1}{2}}$ by $\frac{2/3}{3/4}$
175. Add: $16 1/16; $9 1/8; $5 7/16; $2 13/16.
176. From—$32.31¼
 take $15.12½
177. What is the cost of 17 2/3 dozen of books at $3.75 per dozen?
178. 1/40 bu. is what fraction of a quart?
179. 1/28 lb. is what fraction of an ounce?
180. 1/1584 day is what fraction of a minute?
181. 4/5 qt. is what fraction of a bushel?
182. 4/9 lb. is what fraction of a ton?
183. ¾ in. is what fraction of a rod?
184. Add 2/3 day and 3/4 hour.
185. From $5/8 subtract $3/40.
186. From ⅜ lb. subtract ⅞ oz.
187. Express 26/100 decimally.
188. Express eighty thousand and six ten-millionths decimally.
189. Express 202 ten-billionths decimally.
190. Change .91 to a common fraction.
191. Change .513 to a common fraction.
192. Change 1.36 to a common fraction.
193. Express 4/5 in decimal form.
194. Write 7/25 in decimal form.
195. Write 5/6 in decimal form.
196. Add: 6.61; 636.1; 6516.14; 67.1234; and 5.1233.
197. Add: 4 and 8 tenths; 43 and 31 hundredths; 74 and 19 thousandths; 11 and 204 thousandths.
198. Add: 35 ten-thousandths; .00035; 35 millionths; and 35 ten-millionths.
199. From 97.5168 subtract 38.25942.
200. From .0142 subtract .005.
201. From 1 subtract 1 millionth.
202. Multiply .19 by .125.
203. Multiply 1.325 by .0716.
204. Multiply 43 thousandths by 21 ten-thousandths.
205. Divide .96 by .24.
206. Divide 102048 by .3189.
207. Divide 1.5 by 7.
208. Reduce .035 pk. to pts.

209. Reduce .005 yd. to inches.
210. Reduce .0054875 A. to sq. rd.
211. Express .75 yd. in whole numbers.
212. Express .44 cwt. in whole numbers.
213. Express .3375 A. in whole numbers.
214. 6¼ is what part of 100?
215. 16 2/3 is what part of 100?
216. 87½ is what part of 100?
217. Reduce .72 qt. to bushels.
218. Reduce .77 yd. to miles.
219. At $4.32 per acre how much land can be bought for $59,265?
220. How many miles are there in 20 kilometers?
221. How many Hektoliters in 42 bushels?
222. Reduce 3.8 Ars. to sq. meters.
223. What is 5 1/3% of $150?
224. What is ½% of 456 sheep?
225. What is 1000% of 24.75?
226. 3 sheep is what % of 15 sheep?
227. What % of 9 is 9?
228. What per cent of 2/3 is 2/15?
229. 60 horses are 20% of what number of horses?
230. ¾ is 250% of what number?
231. $85 is 125% of how many dollars?
232. $721 is 3% greater than how many dollars?
233. What number increased by 25% of itself, amounts to 2125?
234. 68 is 66% less than what number?
235. An agent whose commission is 5% receives how much upon a sale of goods amounting to $240?
236. What are the net proceeds on a sale of goods amounting to $180 at 4% commission?
237. An agent receives $1323.54 to cover cost of goods and commission at 8%; what is his commission?
238. A bill of goods amounted to $225.50 list; 20% off being allowed; what was paid for the goods?
239. A bill of goods cost $390.45 at 25 and 5 off; what was the list price?
240. I bought 100 doz. stay bindings at 60c per doz. for 40, 10, 7½% off; what did I pay for them?
241. At ¼% brokerage a broker received $10 for making an investment in bank stock. How many shares did he buy?
242. The brokerage on 95 shares of stock is $11.87½. What is the per cent?
243. I own 35 shares of bank stock; if the bank declares a dividend of 4%, what will I receive?
244. If I invest $39,900 in 6% bonds, at par, what will my income be?
245. What is a man's income who owns 20 1000 dollar U. S. bonds, when gold is 107?
246. What must be paid for 6% bonds to realize an income of 8%?
247. Find the interest on $360 for 20 days at 6%.
248. Find the amount of $430 for 19 days at 4½%.
249. Find the amount of $1000 for 2 years, 3 mon. and 5 days at 9%.

250. I lent $200 at 6% and received $36 interest; how long was the money lent?
251. In what time, at 6% will any principal be doubled?
252. The interest on $400 at 7% was $68.60; how long was the money loaned?
253. I borrowed $600 for 2 yr. and paid $48 interest; what was the per cent of interest?
254. A broker paid $200 interest for the use of $1000 for 2 yrs. 6 mo.; what was the per cent?
255. The amount of $250 for 2 yr. 4 mo. 24 da. was $310, what was the per cent?
256. The interest for 3 yr. at 5% is $341.25; what is the principal?
257. The interest for 1 yr. 4 mo. at 6% is $2.26; what is the principal?
258. The interest for 5 yr. 8 mo. 24 da. at 6% is $28.38; what is the principal?
259. What principal in 9 yr. at 5% will amount to $435?
260. The amount for 2 yr. 6 mo. at 6% is $690. What is the interest?
261. The amount for 4 yr. 3 mo. 27 da. at 4% is $914.94. What is the interest?
262. Find the amount at 6% compound interest of $500 for 3 yr.
263. Find the amount of $600 for 2 yr. at 20% compound interest, payable quarterly.
264. What is the difference between simple interest and compound interest on $1000 for 4 yr. 8 mo. at 7%?
265. What is the interest on $200 for 1 year at 8%?
266. What is the interest on $65 for 4 years at 5%?
267. Find the amount of $7596 for 10 months at 7%.
268. No interest having been paid, find the amount due in 3 yr. on a note for $800 with interest at 8% payable annually.
269. What is the difference between simple and annual interest on $1000 for 5 yr. at 6%?
270. What is the difference between compound interest and simple interest on $5325 for 5 years 6 months at 7% payable semi-annually?
271. Find the date when due, bank discount, and proceeds of a note of $120, dated Oct. 12, payable in 30 days, discounted at 8%.
272. Find the date when due, time of discount, bank discount, and proceeds of a note of $2400 dated Oct. 16, 1876, due Jan. 1st, 1878; with interest at 8% discounted July 26, 1877 at 10%.
273. The proceeds of a note discounted at a bank for 60 days, at 6% were $197.90. What was the face of the note?
274. Find the present worth and discount at 6% of a note of $224 due in 2 years.
275. Find the present worth and discount at 6% of a debt of $675, due in 5 yr. 10 mo.
276. Find the present worth and discount for 5 mo. at 10% of an account for $368.75.
277. What is the cost of a sight draft on New York for $1400 at ½% premium?
278. What will a bill on London for 890£ 8 shillings cost, exchange being at $4.86?
279. How large a bill on Paris can be bought for $1657.60, exchange being at 5 francs 15 centimes to $1.00?
280. What is the ratio of 15 to 20?
281. 4 is the ratio of what number to 13?

282. 2¾ is the ratio of 23⅜ to what number?
283. 2 is to 8 as 6 is to what?
284. What is to 8 as 6 is to 16?
285. 3 is to 7 as what is to 14?
286. 5 is to what as 6 is to 12?
287. If 3 men can dig a cellar in 12 days; how many men will dig it in 6 days?
288. If 3 lb. 12 oz. of tea cost $3.50, what will 11 lb. 4 oz. cost?
289. If 6 men spend $150 in 8 mo. how much will 15 men spend in 20 months?
290. To carry 12 cwt. 75 lb. 400 mi. cost $57.12; what will it cost to carry 10 T. 75 miles?
291. If $100 gain $600 in 12 mo. what sum will $75 gain in 9 months?
292. A and B form a partnership with a capital of $800; A's part is $300; B's $500. They gain $232. What is the share of each?
293. Divide $90.00 among 3 persons, so that the parts shall be to each other as 1, 3 and 5.
294. Divide $692.23 into 3 parts that shall be to each other as 1/3, 3/5 and ½.
295. A put in trade $50 for 4 mo.; B, $60 for 5 mo.; they gained $24. What was each man's share?
296. A and B hire a pasture for $54. A pastures 23 horses 27 days; B, 21 horses 39 days. What will each pay?
297. A put in $300 for 5 mo.; B $400 for 8 mo.; C $500 for 3 mo. They lost $100. Find each one's loss.
298. Mix 6 lb. of sugar at 5c a lb. with 4 lb. at 8c a lb. What will 1 lb. of the mixture be worth?
299. W mixes 3 gal. water with 12 gal. wine at 50c a gal. What is 1 gal. of the mixture worth?
300. I have 30 sheep; 10 are worth $3 each; 12 $4 each; the rest $9 each. Find the average value.
301. Find the square of 65.
302. Find the fourth power of 12.
303. Find the cube of ¾.
304. Find the square root of 529.
305. Find the square root of 915.0625.
306. Find the square root of 384 4/7.
307. The base and perpendicular of a right-angled triangle are 30 and 40. What is the hypotenuse?
308. The hypotenuse of a right-angled triangle is 100, and the base 60; what is the perpendicular?
309. A room is 20 ft. long, 16 ft. wide, and 12 ft. high. What is the distance from one of the lower corners to the opposite upper corner?
310. The surface of a square table contains 8 sq. ft. 4 sq. in. What is the length of one side?
311. The area of a circle is 4096 sq. yd. What is the side of a square of equal area?
312. What is the length of one side of a square lot containing 1 acre?
313. Extract the cube root of 91125.
314. Extract the cube root of 53.157376.
315. Extract the cube root of 9 1/6.
316. A cubical box contains 512 half inch cubes. What are the dimensions?

317. A cubical excavation contains 450 cu. yd. 17 cu. ft. What are the dimensions?
318. The side of a cubical vessel is 1 ft.; find the side of another cubical vessel that shall contain 3 times as much.
319. How many square feet in a floor 17 ft. long and 15 ft. wide?
320. A room is 10 ft. long; how wide must it be to contain 80 sq. ft.?
321. How many yards of flannel ¾ yd. wide will cover a floor 18 ft. long and 15 ft. wide?
322. The base of a triangle is 15 ft. and its altitude 12 ft. What is its area?
323. Find the area of a triangle whose sides are 13, 14 and 15 ft.
324. Find the area of a field in the form of a trapezium of which a diagonal is 50 rd. and the perpendicular to the diagonal from the opposite angles 30 rd. and 20 rd.
325. The diameter of a circle is 48 ft. What is the circumference?
326. Find the area of a circle whose radius is 21 ft.
327. The area of a circle is 6 sq. ft. 98.115 sq. in. What are its diameter and circumference?
328. Find the surface of a cube each side being 37 in.
329. Find the surface of a cylinder, its altitude being 5 ft. and radius of the base 2 ft.
330. How many cubic yards in a room 24 ft. long, 18 ft. 6 in. wide and 10 ft. 7 in. high?
331. What is the convex surface of a cone of which the slant height is 25 ft., and the diameter of the base 8 ft. 6 in.?
332. Find the volume of a square pyramid of which each side of the base is 5 ft. and altitude 21 ft.
333. A square pyramid is 477 ft. high; each side of the base is 720 ft. How many cu. yd. in the pyramid?
334. What is the surface of a sphere of which the diameter is 1 ft.?
335. Find the volume of a sphere 13 ft. in diameter.
336. The volume of a sphere is 1 cu. ft. What is its diameter?
337. What is the 99th term of a decreasing series in which the first term is 329 and common difference is ⅞?
338. The first term of a series is 2, the last 20, and the number of terms 7. What is the common difference?
339. The extremes in a progression are 2 and 50; the number of terms 24; find the sum of the series.
340. The first term of a decreasing series is 262144; the ratio 4; the number of terms 9. Find the last term.
341. Find the sum of the infinite series 1/3, 1/9, 1/27, etc.
342. Find the sum of an infinite series of which the greatest term is .3 and the ratio 10.
343. Find the yearly premium which a man of 25 must pay for a whole life policy for $5,000, at $20.50 per $1,000 per year.
344. A house was insured for $12,000 at ½% premium. After paying four yearly dividends, the house valued at $15,000 is destroyed by fire. What is the loss counting the premiums?
345. The revenue tax on a certain article is $1.10. How much will be added to the consumers' price if successive profits of 15%, 15%, and 25% to be paid?

# APPENDIX B

## "ANSWERAIRE" B

January, 1927

DEAR PARENTS:

We are trying to find out just what topics in arithmetic you and other adults use—or have had to use—since you left school. We do not want to know the kind nor amount of arithmetic you use in your business, but just what you use in your life ASIDE from your business. We are trying to find out whether the arithmetic that was—or is now—taught in the public schools is too much or not enough, and what kinds should be taught. Will you assist us by filling out the accompanying blank to the best of your ability and returning it immediately by your child to the teacher?

*First*—You will notice on the left margin of the blank a list of problems which are given as representative types to indicate the kind of arithmetic you *do* or *do not* use in your everyday life. You will notice across the top of each sheet the seven headings: BUYING AT STORE (Groceries or Dry-goods), MAKING CHANGE, READING NEWSPAPERS OR MAGAZINES, INVESTING YOUR SAVINGS, WRITING LETTERS, ETC., FILLING OUT INCOME TAX SHEETS, and TRAVELING.

*Second*—Read over the problems, one at a time, and make a check under "YES" or "NO" under each of the headings at the top of the page. *Be sure not to miss checking each problem either "Yes" or "No" under each of the headings.*

EXAMPLE:

| ASIDE FROM BUSINESS OR VOCATIONAL NEEDS, HAVE YOU EVER HAD OCCASION TO USE PROBLEMS LIKE THESE SINCE YOU LEFT SCHOOL? (Check each problem under "Yes" or "No" under each of the headings to the right) | BUYING AT STORE | | MAKING CHANGE | | READING PAPER OR MAGAZINES | | INVESTING YOUR SAVINGS | | WRITING LETTERS, ETC. | | INCOME TAX SHEETS | | TRAVELING | |
|---|---|---|---|---|---|---|---|---|---|---|---|---|---|---|
| | YES | NO | YES | NO | YES | NO | YES | NO | YES | NO | YES | NO | YES | NO |
| 75. Add: ½ and ¾ | | | | | | | | | | | | | | |

*Third*—You are *not* to solve any of the problems, but check whether you have ever used the kind of arithmetic *needed* to solve them. Then fill out the following blanks:

Your Name ______________________________ Age __________

Do you support a family? ______ Grade at which you left school ______

CHECK UNDER "YES" OR "NO" WHETHER YOU HAVE EVER USED PROBLEMS LIKE THESE. SOLVE NO PROBLEMS. RETURN ALL SHEETS TO TEACHER IMMEDIATELY

| ASIDE FROM BUSINESS OR VOCATIONAL NEEDS, HAVE YOU EVER HAD OCCASION TO USE PROBLEMS LIKE THESE SINCE YOU LEFT SCHOOL? (Check each problem under "Yes" or "No" under each of the headings to the right) | BUYING AT STORE | | MAKING CHANGE | | READING PAPER OR MAGAZINES | | INVESTING YOUR SAVINGS | | WRITING LETTERS, ETC. | | INCOME TAX SHEETS | | TRAVELING | |
|---|---|---|---|---|---|---|---|---|---|---|---|---|---|---|
| | YES | NO | YES | NO | YES | NO | YES | NO | YES | NO | YES | NO | YES | NO |
| 2. Express the following in figures: Thirty-five billion. | | | | | | | | | | | | | | |
| 5. Read the following: $320,803,538.00. | | | | | | | | | | | | | | |
| 7. XV. | | | | | | | | | | | | | | |
| 10. Reduce $3.00 to mills. | | | | | | | | | | | | | | |
| 13. At $1.75 per yard what will nine inches of silk cost? | | | | | | | | | | | | | | |
| 14. At 25c per foot what will three rods of wire fence cost? | | | | | | | | | | | | | | |
| 18. What is the value of the crown in U. S. money and in all countries where the crown is used? | | | | | | | | | | | | | | |
| 21. What is the value of the Peso in U. S. money? | | | | | | | | | | | | | | |
| 22. What is the value in U. S. money of the English pound sterling? The shilling? | | | | | | | | | | | | | | |
| 23. What is the value in U. S. money of the franc, centime, etc.? | | | | | | | | | | | | | | |
| 24. Do you use the German mark, pfennig, thaler? | | | | | | | | | | | | | | |

Occupation ________________________________________
Name of Newspapers and Magazines you read:

1. ________________________________________
2. ________________________________________
3. ________________________________________
4. ________________________________________

Do you ever check your monthly bank statements? ________ Do you ever check your grocery or drygoods bills, or other bills when they are presented to you to see if they are correct? ________

*Return all blanks to the teacher immediately, please.*

*Thank you.*

Appendix B is made up from problems selected from Appendix A. They are as follows:

Problems numbered, 2, 5, 7, 10, 13, 14, 18, 21, 22, 23, 24, 25, 26, 27, 30, 31, 32, 35, 37, 38, 40, 43, 47, 54, 56, 57, 58, 61, 67, 73, 76, 79, 82, 88, 91, 97, 100, 104, 108, 109, 115, 116, 118, 121, 124, 126, 127, 129, 133, 139, 142, 145, 150, 153, 154, 157, 162, 163, 169, 170, 172, 176, 181, 186, 187, 190, 193, 195, 200, 202, 205, 208, 209, 211, 214, 215, 216, 217, 218, 220, 223, 226, 230, 235, 238, 239, 240, 242, 246, 248, 250, 253, 256, 260, 262, 265, 269, 271, 274, 277, 278, 280, 283, 284, 289, 293, 295, 298, 301, 302, 303, 305, 311, 313, 316, 319, 323, 325, 326, 330, 341, 343, 344.